Woody Ornamentals
for Deep South Gardens

Woody Ornamentals for Deep South Gardens

*David J. Rogers
and Constance Rogers*

Foreword by Ghillean T. Prance
Drawings by Mitzie Briscoe Edwards
Photographs by David J. Rogers

University of West Florida Press / Pensacola

Copyright 1991 by the Board of Regents of the State of Florida

Printed in the U.S.A.

The University of West Florida Press is a member of University Presses of Florida, the scholarly publishing agency of the State University System of Florida. Books are selected for publication by faculty editorial committees at each of Florida's nine public universities: Florida A&M University (Tallahassee), Florida Atlantic University (Boca Raton), Florida International University (Miami), Florida State University (Tallahassee), University of Central Florida (Orlando), University of Florida (Gainesville), University of North Florida (Jacksonville), University of South Florida (Tampa), University of West Florida (Pensacola).

Orders for books published by all member presses of University Presses of Florida should be addressed to University Presses of Florida, 15 NW 15th Street, Gainesville, FL 32611.

Library of Congress Cataloging-in-Publication Data

Rogers, David J. (David James), 1918–
 Woody ornamentals for deep south gardens / David J. Rogers and Constance Rogers; foreword by Ghillean T. Prance; drawings by Mitzie Briscoe Edwards; photographs by David J. Rogers.
 p. cm.
Includes bibliographical references and indexes.
ISBN 0–8130–1011–X (alk. paper)—ISBN 0–8130–1021–7 (pbk.: alk. paper)
1. Ornamental woody plants—Southern States. 2. Landscape gardening—Southern States. I. Rogers, Constance. II. Title.
SB435.52.S67R64 1991   90-47372
713'.0976—dc20   CIP

Contents

Foreword, Ghillean T. Prance vii

About the Authors ix

Introduction xi

How to Use This Book xvii

Deep South Gardens 1

Table of Horticultural Characteristics and Landscape Planning Aids 233

Selected Reading List 269

Index of Common Names 271

Index of Scientific Names 285

Index of Families, Genera, and Species 292

Foreword

The Deep South, familiar to American gardeners as Zone 8 of the plant growth zones, offers the opportunity to grow a wide range of plants. Here we have a description of large numbers of subtropical trees and shrubs that cannot be grown in more northerly gardens. The plants included come from around the world, but it is significant that this region has also made its own contribution to the gardens of other parts of the world. In this guide to the woody ornamental plants of the Deep South, I find many familiar American plants that are grown in Europe, such as *Franklinia* or the magnificent southern magnolia that grows against one of the Museum buildings of the Royal Botanic Gardens, Kew.

This book is full of useful information about all the commonest trees, shrubs, and vines that are cultivated in the South, as well as some of the rarer species. People who are planning what to plant in their gardens will find the lengthy table of horticultural characteristics a considerable help for their landscaping. A guide of this sort is always more useful when it is well illustrated so that one can recognize the plants described, and the drawings by Mitzie Briscoe Edwards are most welcome.

David Rogers has had a long association with the study of useful plants, and it is certainly fortunate for the many gardeners of the Deep South that he has returned to his native habitat and devoted many hours of his retirement to the gathering of data for this volume. Those of us who live farther away from that region will also appreciate the information about many of the plants that adorn our gardens.

I hope that you will not only enjoy this book but that it will lead you to plant and enjoy many of the trees and shrubs described, or at least look out for them in other public and private gardens.

Ghillean T. Prance
Director
Royal Botanic Gardens, Kew

About the Authors

David Rogers. Born in De Funiak Springs, Florida. Completed high school at Walton County High. Received B.A. in botany and horticulture from the University of Florida, Gainesville. After five years of military service in World War II, received Ph.D. in systematic botany from Washington University, St. Louis, Missouri. Taught botany and related biology courses at Allegheny College, Meadville, Pennsylvania. Became Curator of Economic Botany at the New York Botanical Garden, New York City, a post that involved researching plants of economic importance and editing a scientific journal, *Economic Botany*. Became Professor of Botany at Colorado State University and subsequently at the University of Colorado, Boulder. Retired from the University of Colorado as Emeritus Professor and moved back to De Funiak Springs in 1980. Traveled throughout his career to research botany in many parts of the world, including Latin America, Turkey, tropical west Africa, and several European countries. Served as a consultant for several agencies of the United Nations, the United States Agency for International Development, the Rockefeller Foundation, and several private corporations in the United States.

Constance Rogers. Born in Bridgeport, Connecticut. Finished high school in Syracuse, New York. Received R.N. at the School of

Nursing at Endicott Johnson Memorial Hospital, Johnson City, New York. Served in the United States Army Nursing Corps during World War II, with overseas service in Europe. Married David and raised three children. Taught in the School of Nursing at Meadville City Hospital, Meadville, Pennsylvania. Worked many years as a volunteer for the American Red Cross and several other social agencies. Resumed work as a nurse for a time after children were grown. Studied botany at the University of Colorado, then worked with David in his laboratory for ten years before retirement.

Introduction

The plants included in this book are the trees, shrubs, and woody vines that the authors have seen growing in the Deep South "Zone 8." Readers familiar with garden books and plant catalogs are familiar with Zone 8, which designates a latitudinal area of the United States with more or less the same average minimum temperatures of 20 to 30 degrees Fahrenheit. Our efforts have been concentrated in the central part of the zone, with only a few forays into the eastern and western ends. While the definition of Zone 8 includes much of lower Texas as well as southern California, we think that the area of greatest similarity ends with the eastern most side of Texas, not far west of Galveston.

Returning to Zone 8 in 1980 after many years absence, we noticed that many new cultivated plants we did not know had been introduced and planted. There has been a great surge in population since World War II and, with it, the construction of many new homes. Most of these new homes were financed either through the GI Bill or through the FHA, and both agencies required that any new home must have a certain, minimum amount of landscaping. The requirement was a tremendous boost to nurseries, wholesale and retail, and they responded by making available to the buying public a large number of introduced species and new variations of

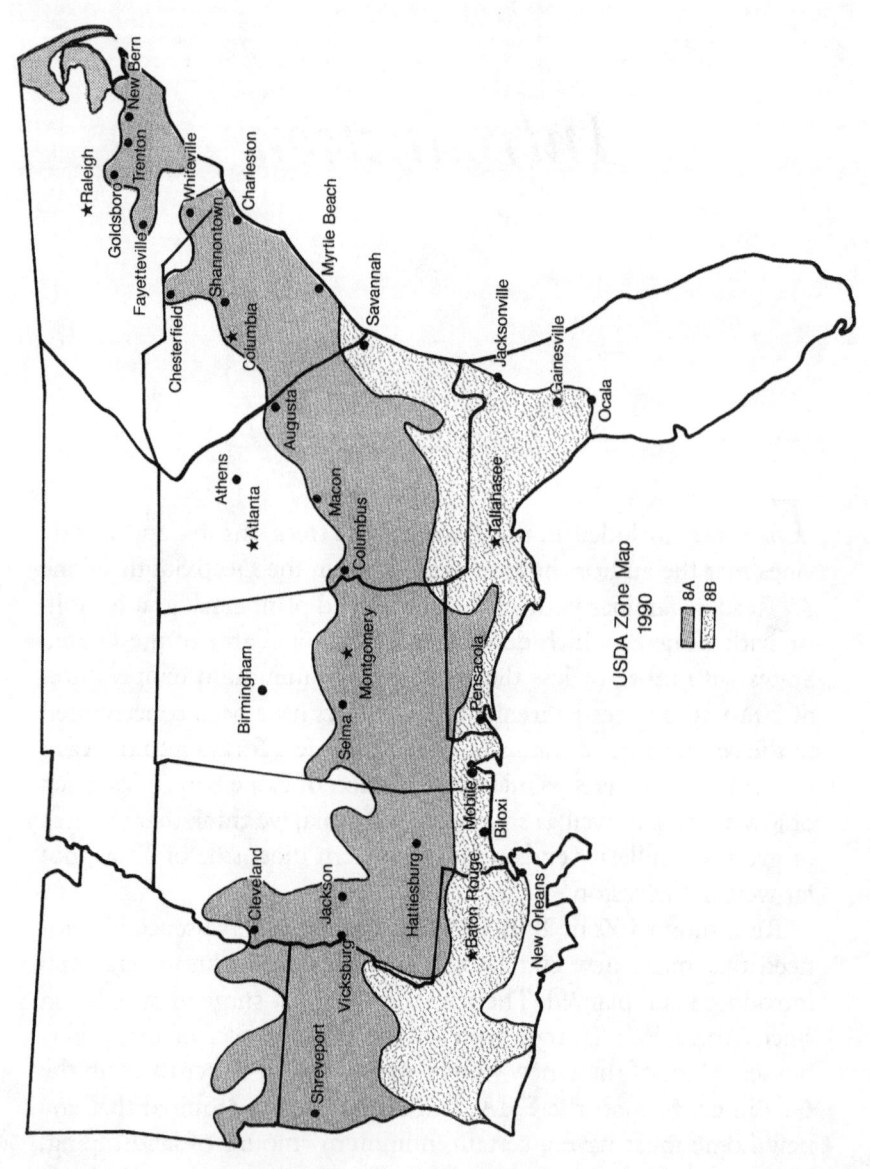

The area covered by this book, the Deep South, corresponds to Zone 8 on the Plant Hardiness Zone Map, the January 1990 edition, released by the United States Department of Agriculture. People familiar with earlier hardiness zone maps will recall that we were placed in Zone 9 in the 1965 edition. The current map results from correlations made by computer from 14,000 reporting weather stations for each year since 1965. The resulting map much more accurately represents the boundaries of the zones, which are areas with average minimum temperatures of a ten degree range, in our case, between 10 and 20 degrees Fahrenheit. The contiguous 48 States of the United States are divided into ten zones. The coldest zone, 1, is in interior Alaska. The warmest zone, 11, is found only in extreme south Florida and the warmest coastal regions of southern California.

You will note that each zone is divided into two subzones, a and b, representing a 5-degree difference in minimum temperatures. This gives further refinement to the map, and is useful in deciding whether or not a particular plant has a better chance of survival in a particular locality. Additional factors to temperature play a role in success or failure of a plant to survive in a particular locality within zones. Factors such as moisture, soils, exposure to sun, and perhaps others, may also play a role in the chance of survival of any particular species or variety of plant. Different species vary greatly in their tolerance of minimal temperatures so that many of the species in Zone 8 may grow much farther north, or south, of this zone. For this reason, this book may be useful in parts of Zone 7 and/or Zone 9. We limit the western boundary of coverage to the Texas-Louisiana border.

old species. Many of the introductions are woody plants, trees, shrubs, and woody vines.

Since we wanted to use plants in our landscaping that would have a good chance of survival, we decided to make an informal survey of what is growing in this area. Our "sampling technique" was to drive around in various communities and note what we found. If something caught our eye that we didn't know, we stopped to inquire about it or to take a specimen for identification if we couldn't determine what it was upon examination. (We were very pleased when we asked the owners to share information on the plants we stopped to see: without exception, perfect strangers were generous with information and with sample material to be used in identification.) We did not rely on our own sampling entirely, but also enlisted the aid of the members of the Garden Club of De Funiak, who enthusiastically agreed to help complete our survey.

The native flora has contributed a rather large number of very handsome plants and much beauty to our yards and gardens. But many of these plants are either endangered species or are listed as rarities that must not be taken from their native locations. *We universally recommend that no plant be taken from the wild. Instead, anyone interested should contact various nurseries to get plants that have been propagated from wild plants, or that have been in cultivation long enough so that a good supply is available.* Without exception, nursery-grown plants have a much better chance of survival in your own garden or yard than specimens collected in the wild because their root systems are more confined and transplanting is much easier. Our native plant areas are being lost at an alarming rate, and we all must share responsibility for preventing their further reduction.

One word of caution: we have not identified all of the garden "cultivars," or cultivated varieties, of many of the species included in this work. Some, such as *Camellia japonica*, have a very large number of cultivars, and we could not possibly know which of several hundred different names refers to any one particular variety. There are a few specialists for each group of plants whose major endeavor is to keep up with the named cultivars. Often, there are also societies for each important group, and these have compilations at their headquarters dealing with the named cultivars.

We have relied on the knowledge and personal assistance rendered by Dr. Frederick G. Meyer, Supervisory Botanist in charge of the Herbarium at The National Arboretum, Washington, D.C., for plants we could not identify ourselves. He has taken much time out of his busy schedule to review the manuscript, giving many important suggestions and changes. We thank Dr. David Hall and his assistants at the University of Florida, Gainesville, where we received much expert assistance in naming some of our plants. We thank the members of the De Funiak Springs Garden Club for their enthusiastic support, specifically Mrs. Nell Sawyer, Miss Mary Green, and Miss Mary Elliot, who served on a committee to advise and assist us. Dr. Fannie Fern Davis, formerly of Fort Walton Beach, gave us many valuable suggestions and corrections. Dr. James Servies, Emeritus Director, the Pace Library of the University of West Florida, kept us from many grammatical errors and excessive verbiage, thereby making the text more readable and enjoyable. Mr. Jackie Balkcom, County Forester, Walton County, Florida, not only collected material for illustration of the pines and other tree species, but also read the section on the pines for adequacy and accuracy. Mr. and Mrs. Ray Spikes, owners of Spikes Nursery in De Funiak, read the whole manuscript, offered many valuable suggestions, and generously allowed us to take specimens for illustration when we had difficulty locating plant material.

We are indebted to the book *Hortus III* (New York: Macmillan, 1976), compiled by the staff of the Bailey Hortorium at Cornell University, for the scientific names and other types of information that we did not have at our disposal. Other publications we used are listed in the "selected reading" list.

Illustrations of the following species were redrawn from the following publications: Fig. 106, *Osmanthus americanus,* from N. L. Britton, *North American Trees,* (New York: Holt, Rinehart and Winston, 1908), p. 814; Fig. 110, *Paulownia tomentosa,* from Britton and Brown, *An Illustrated Flora of the Northern United States and Canada,* 2nd ed., III (New York: Dover Publications, orig. 1913),

p. 189; and Fig. 159, *Ternstroemia gymnanthera,* from Richard H. Beddome, *Flora Sylvatica for Southern India* I (Madras: Gantz Brothers, 1871), table 91.

We would be the first to admit that the list of plants included is probably incomplete. We hope that those who read this book will report to us any of their favorite plants which we have omitted. We are solely responsible for any errors included.

How to Use This Book

This book is arranged alphabetically according to plants' scientific names. Though this organization may appear academic or excessively technical, there is no better way to avoid having to repeat some plant names several times in the text. The International Rules of Botanical Nomenclature specify the use of Latin for the scientific names of plants. Thus, all nations—whether using the Roman alphabet, Arabic, Cyrillic, Chinese, or Japanese—must use Latin names. All the plants we deal with have, by international rules, two names: *genus* (plural *genera*) and *species* (the same spelling in both singular and plural). An example is *Rosa alba*, which translates as white rose. Note the use of *italics* for the scientific names, a universal practice whenever the scientific name is written. No other species of the genus *Rosa* can be called by the same species name, but the word *alba* can be used in combination with another, different, genus name.

Many people designate plants using common names; thus an index to the common names is provided. With each common name in the index you will find a reference to the scientific name, and since the *generic* scientific names are arranged alphabetically throughout the text, you can go directly to the plant in question. Only the genus name is alphabetized; if there is more than one species in the genus used in cultivation here, then all the species will be found under the heading of their shared genus name. If there are several common names for the same plant (a frequent occurrence), each common name is listed in the index and the proper scientific name referred to. Common names that refer to one scientific name are listed in small capitals just after the scientific name in the text.

To facilitate location of names, the following format is used: the genus name in boldface, capital letters is followed by general descriptions and comments. After the generic descriptions, the scientific names of the species are given at the beginning of separate paragraphs. The scientific name is italicized (for example, *Hibiscus mutabilis*) and is followed by the common name (in this case, Confederate rose). A description giving more detail on the species discussed follows the common name.

Many plants either have no common name or else have a common name that is the same as the scientific name. For example, the genus name *Forsythia* is the scientific name given to a particular group of plants. While these plants also have a common name, golden bells, it is less frequently used and most folks use the proper scientific name. In such a case as this, the reader should go directly to the genus, sequenced alphabetically in the text.

Three indices are provided: an index of common names to facilitate finding a plant in the text, and two additional indices of scientific names.

The first of the scientific indices is arranged by genera and species, giving the family name in each case. The authority for the species name is also given in this index. The second scientific index is also of scientific names, arranged with family names first and then the genera and species mentioned in the text that belong to these families. If you are interested in other species in the book that are related to camellia, for example, look in the first scientific index to find the family name, then in the second index, under the family name, to find other plants mentioned in the text that are related to the camellia.

We use two symbols in the text to give some additional information about the species, when appropriate. You will find an example of the first of these symbols in the text in the first genus included, *Abelia*. Between the genus and the species name, an × appears, *Abe-*

lia × *grandiflora*. The × is shorthand used by botanists and horticulturists to indicate that the species is actually a hybrid between two other species, but a hybrid that has been found to demonstrate a stability of characteristics generation after generation. The second symbol found infrequently in the text is the question mark in parentheses (?) just after the species name. An example is found in *Eucalyptus, Eucalyptus polyanthemos* (?). For some reason, the species name is in doubt, either because we are not sure that this is the correct name, or because we may think that the species should be identified as another species. Sometimes, we cannot be sure whether the name we have used is correct or incorrect. If you are curious why this can happen, consider the fact that the genus *Eucalyptus* has over 600 species, and the distinction between the individual species rests on characteristics which are not exhibited by the plants we have been discussing. We do the best we can, and the question mark warns you that we may have given the wrong name.

Last, there is a large table for all of the included plants, arranged alphabetically by common names, in sections by their heights—tall, medium, and small trees, shrubs, woody vines, and ground covers—giving characteristics important in determining their horticultural value. Here you will find comparative characteristics which provide a useful tool for landscape design. For example, a shrub that will thrive in a shady area is aucuba (*Aucuba japonica*). If a gardener wants shade in summer but sun in winter, a number of small, medium, or large trees will meet that particular need. This table frees the gardener from total dependence on a landscape architect who may (or may not!) know the gardener's requirements and preferences.

Deep South Gardens

ABELIA. The abelia is one of the oldest plants used in gardens in the Deep South (Zone 8). It is a very satisfactory shrub, growing to about eight feet tall, and used in a variety of conditions from partial shade to full sun. Although an undemanding plant, it will respond to good growing practices, such as a little fertilizer, either inorganic or organic, in spring and summer. If the plants become leggy, they may be pruned in almost any season, but it is best to do so in the dormant season, January or February.

―――*Abelia* × *grandiflora* (fig. 1). There really is no common name for this plant, and people just know it as "abelia." The abelia is a summer-flowering plant, and clusters of small, bell-shaped, pinkish-white flowers are produced almost continuously from May to late in the fall. The leaves are glossy and dark green, and the plants are almost evergreen, though many of the leaves drop in winter.

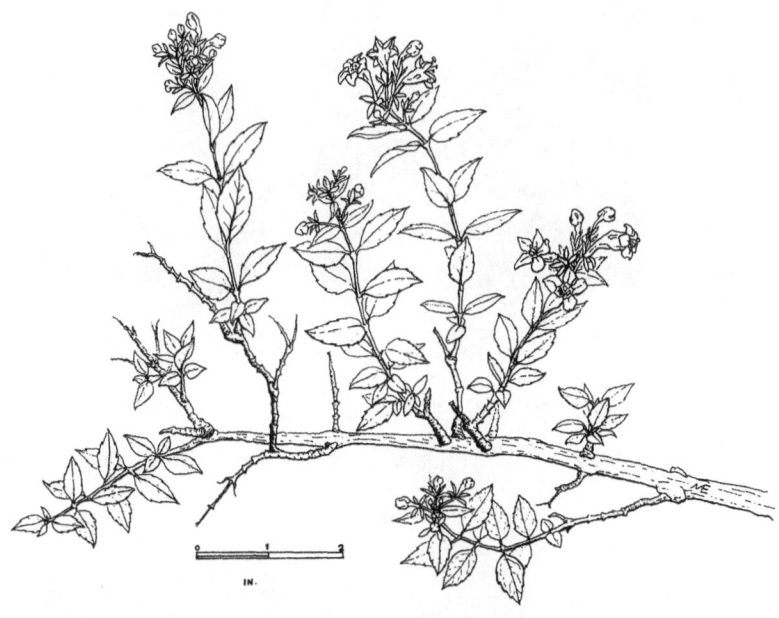

Figure 1. Abelia × *grandiflora* (Abelia)

ACER. The genus, commonly called maple, has a number of useful species in most temperate regions of the northern hemisphere where the plants are found as natives. Many of the species are used commercially, the best known of which is the sugar maple, which not only supplies sap that can be made into syrup and sugar but also provides one of the best furniture woods known. There are both trees and shrubs in the genus, and they are all deciduous. In our region (Zone 8), one native species and one introduced species are frequently used horticulturally.

Figure 2. Acer palmatum (Japanese Maple)

————*Acer palmatum,* JAPANESE MAPLE (fig. 2). Many of our most beautiful plants come from Japan, and the Japanese maple is one of the most frequently planted. This species is a small, graceful tree, growing to about 20 feet tall, frequently branched. The leaves are the predominant horticultural feature, both from the structural and color aspect. They are deeply 5–11 lobed (much more so than the following species), and the margins are prominently serrate or saw-toothed. At maturity, the leaves are red, some lighter, some darker red, depending on the individual variation used. The flowers and fruits are also red, but they bloom and fruit later in the season than the red maple. Culture of this tree is not difficult, requiring only a good planting medium, regular watering, and addition of a balanced fertilizer, such as 8–8–8, with added minor nutrients.

————*Acer rubrum,* RED MAPLE (Southern form) (fig. 3). This native species is one of our earliest harbingers of spring, with bright red clusters of flowers followed by equally bright red clusters of fruits called "keys." Many people confuse this species with the preceding one because of the common name problem. Both species have red pigments, but the native red maple has green leaves, whereas the Japanese maple has red leaves (but only in the cultivated forms—the wild representatives of this species have green leaves). Such is the joy (and confusion) frequently caused by common names. This native tree is about 25 to 50 feet tall and grows most frequently in swampy areas, but it can be grown successfully in upland yards. The leaves are generally three lobed, much less deeply so than the Japanese maple. At maturity, they are a dark, glossy green and, like all maples, deciduous. A tree purchased at a nursery is more reliable than one transplanted directly from a swampy area to the yard; that transition might be too great a shock for the tree. You will save both time and money by using nursery-grown plants.

❀

AESCULUS. This is a small genus of trees and shrubs from North America, Europe, and Asia. The common name is either "horse chestnut" or "buckeye." The first one is descriptive of the outer layer

6 ❦ *Aesculus*

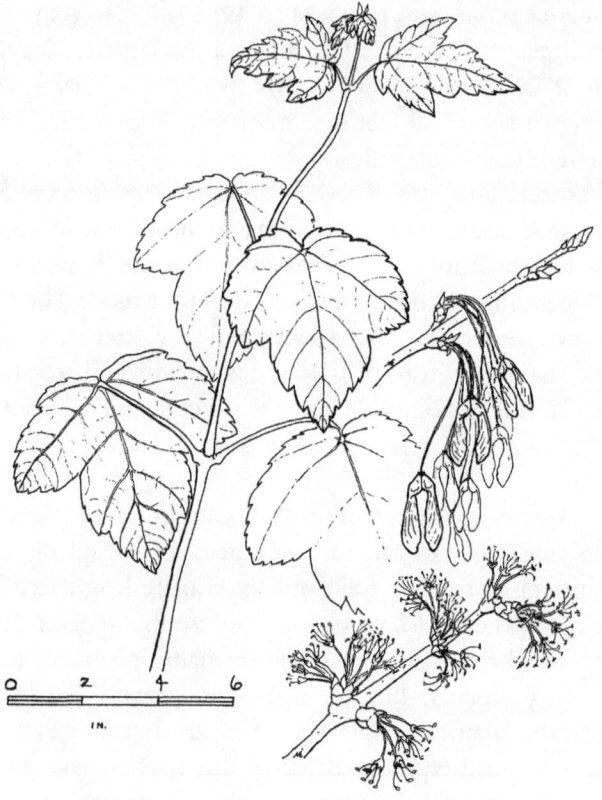

Figure 3. Acer rubrum (Red Maple)

of the fruit, which has prickles somewhat reminiscent of a regular chestnut fruit. The second common name comes from the resemblance of the seed to the eye of a buck deer. The Ohio nickname, the "buckeye," comes from the latter. These plants have compound leaves, that is, several leaflets on one leaf. They are deciduous species, dropping their leaves in the winter, and can be used for horticultural purposes in several different ways. Only one of the species, the following, is found in our part of the country.

———*Aesculus pavia*, RED BUCKEYE (fig. 4). This handsome shrub or small tree, which grows to 10 or 12 feet tall (to 25 feet farther north), is a native that can be used in the garden. Its flowers, in

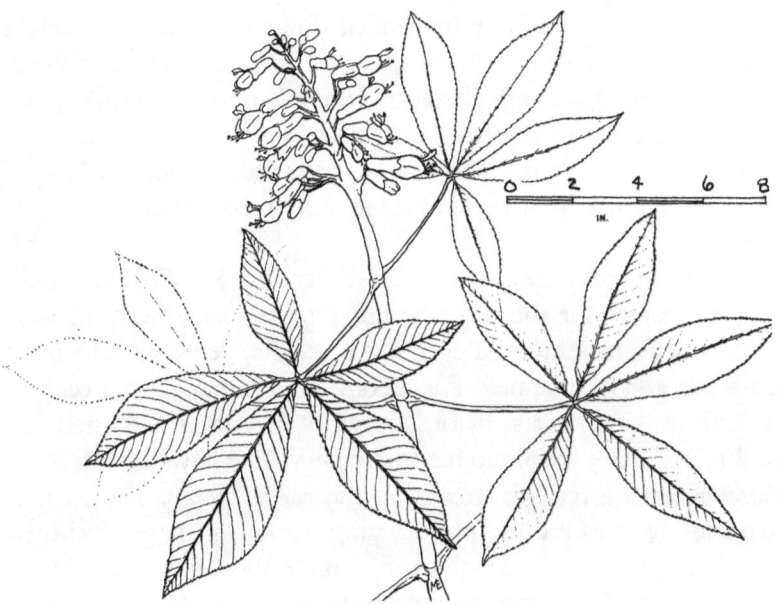

Figure 4. Aesculus pavia (Red Buckeye)

clusters at the end of the branches, make a bright red splash of color from late May to the end of June. Most of the plants we have seen are low shrubs, not more than five feet tall, but one fine plant at Eden State Park, in Point Washington, Florida, is about 12 feet tall. These are good plants to have because they do very well in shade. They do not demand much care.

AGAVE. Strictly speaking, all the members of this genus (some 300 species) are succulents and do not belong with a discussion of woody plants. However, these species generally have a short, thick stem to which the succulent leaves are attached, and this stem is more woody than succulent. The species are natives of the Americas, North and South, but many have been taken to the Old World and, to some extent, have been naturalized there. The species generally occupy desert, or near-desert, habitats. Mexico and the southwestern United States are frequently associated with a drawing showing a man in a large hat pulled down over his face, leaning against a wall, asleep, with an agave in the background, to imply both heat and dryness. Cultivation is simple once the plants are established from transplants or offshoots from a parent plant or, in some cases, from seed. Little water is required, but for any kind of growth a regular application of some water is beneficial.

———*Agave americana*, CENTURY PLANT (fig. 5). We don't know where the idea for the name "century plant" came from, because these succulents seldom, if ever, live to be 100 years old. The plant does live a long time before it blooms, and it may seem a century (usually about 25 years) before it decides to send up a tall stalk, 15 to 20 feet. There are numerous attractive yellow flowers, with some varieties having reddish streaks on the yellow petals. The plant is generally used as a centerpiece setting because its large thickened leaves are produced in a tight spiral on the short stem. The leaves, up to five feet long, may be eight to ten inches across, and four or

Figure 5. Agave americana (Century Plant). Plant 4 ft. tall.

five inches thick, tapering to a sharp spine. They are gray-green, covered with a heavy, waxy epidermis that aids in prevention of water loss from the succulent interior. The plants flower only once, and they die after flowering and fruiting have occurred. We can raise these desert-loving plants in deep sand which, because of its low water retention, is more often dry than wet.

❦

ALBIZIA. The mimosa belongs to this genus of over 100 species, all native to the Old World. Ours came from either Iran, China, or Japan. All species in this genus have feathery, compound leaves in which the leaf blade is divided into a number of leaflets. These plants shed their leaves in winter, and all have flowers like powder puffs. The really showy part of the flower is the large number of brightly colored stamens, as many as 50 or more per flower. The petals of the flower are rather small and not showy. The fruits are pods, much like beans (members of the same family). These are flattened and dry when ripe, and many are formed on each plant each year.

––––––*Albizia julibrissin,* MIMOSA, SILK TREE (fig. 6). This graceful, small tree or large shrub with gently curving gray stems is not only a common sight in our yards and gardens, but is frequently found, escaped, along roadsides and in old fields. This might give the impression that it is a native, but it is not. The plants have been grown successfully in southeastern gardens for a long time. Unfortunately, the trees are rather short lived. The pink to reddish flowers appear from May to midsummer, a time during which many showy trees have already passed their best flowering stages. Cultivation is simple, since the plants require little attention.

❦

ALEURITES. This genus is a member of the spurge family, Euphorbiaceae and, like several other genera in the family, contains poisonous principals that are quite dangerous if not deadly. Tung oil trees were first introduced to this country from China as a source of commercial oils for fine paints and varnishes. The trees were

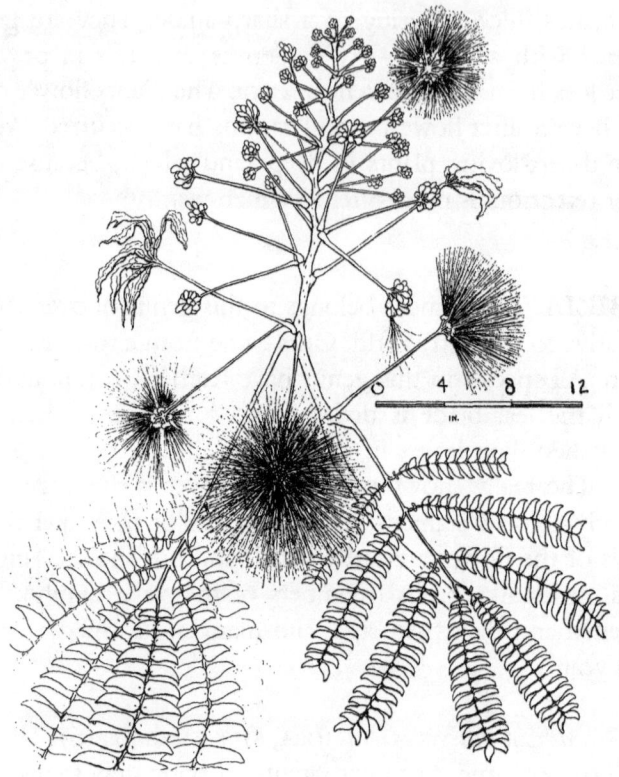

Figure 6. Albizia julibrissin (Mimosa)

found to thrive in the Deep South, and large acreages were planted in the 1920s and 1930s. The harvested seeds are the source of the oil, and these must be crushed and the oil extracted in a particular procedure. All growers had to send their harvest to one seed-crushing mill in Mississippi. Shortly after oil production reached large quantities, a synthetic substitute was found that could be produced more cheaply from other raw materials. Thus the tung oil industry slowly faded away, and the one processing mill closed. It was later found that the synthetic substitute did not have the same qualities as the tung oil, but by then the processing mill had gone out of business and growers had turned their attention to other types of agriculture.

———*Aleurites fordii*, TUNG OIL TREE (fig. 7). The trees in many groves are still growing untended and producing reduced quantities of nuts. The trees produce magnificent white and pink flowers in April, and they are becoming recognized as good decorative plants that can be used in a variety of settings. The deciduous leaves are about the same shape and size as the catalpa (the blade 6–8 inches long, 5–7 inches wide). The trees grow to about 15 or 20 feet tall and are nicely rounded in their crowns. They grow readily from seeds and need little, if any, attention. They seem to have few natural enemies. The seeds (*not* nuts) are poisonous and also the leaves, which some people react to as when exposed to poison ivy.

ARUNDO. This small genus of grasses has only a few species, all of which are rather tall—up to 18 feet—usually growing in clumps or clusters. Only one species is found in cultivation.

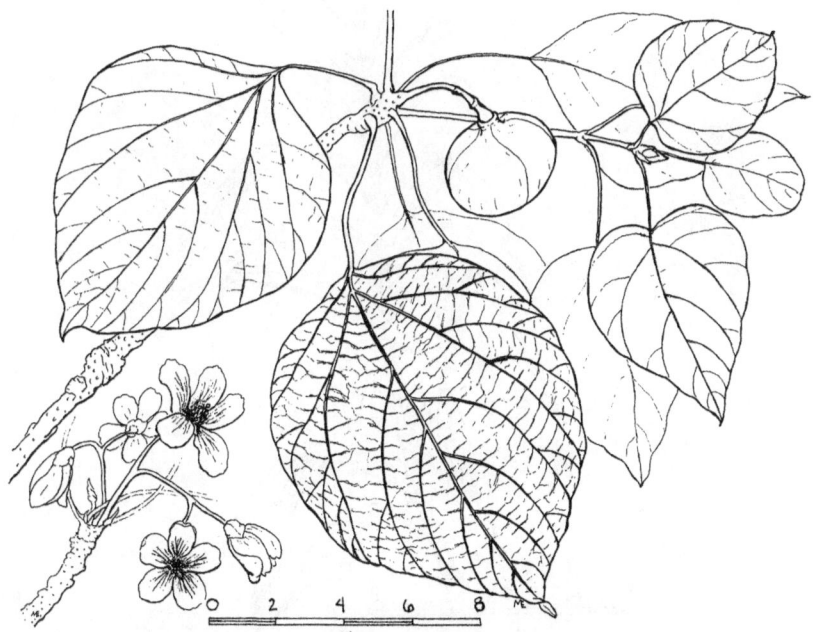

Figure 7. Aleurites fordii (Tung Oil Tree)

———*Arundo donax,* GIANT REED (fig. 8). This tall, graceful plant has strap-shaped leaves that come out of the stem in a single rank, giving the appearance of a palm frond. It seems to have originated in the Tigris and Euphrates valleys of the Middle East. Its first use may have been for making musical instruments, the individual joints cut to differing lengths to make "Pan pipes." Today, sections cut from the dried, mature stems, thinned to a vibrating edge, are used in wind instruments such as the clarinet and saxophone. Used as an ornamental, the giant reed makes a good accent plant placed, for example, on each side of a driveway. There are two varieties that we know about, one with leaves of light green, the other with varie-

Figure 8. Arundo donax (Giant Reed)

gated green and white striped leaves. The plants will grow well in our sandy soils but should be placed in the open sun for best growth.

❧

AUCUBA. Botanists indicate that this small group of evergreen shrubs is related to the dogwoods, in the family Cornaceae. Native from Japan to the Himalayas, they are grown for their showy foliage and sometimes for their bright red fruit. They grow best in shady areas and are easily propagated from stem cuttings.

———*Aucuba japonica,* JAPANESE LAUREL, GOLD DUST SHRUB (fig. 9). This handsome shrub grows to about 15 feet tall, so individuals should not be planted too close together or in front of windows. (Frequently we do not realize the growth potential of plants we bring home from the nursery and do not give them sufficient space when we plant them. The results of such mistakes are seen too often—tangles of indistinguishable plants without any character. A little planning and patience, plus some questions to the nursery attendant on growth potential and speed of growth, will pay off a few years later and will save considerable amounts of time and money.) There are two varieties of this shrub we know about: a form with dark green leaves and a form with variegated, dark green and bright yellow-spotted leaves called gold dust shrub. The latter is most commonly seen.

❧

BETULA. The birches, members of this genus, grow in the Northern Hemisphere from the arctic to the subtropics. They are trees or shrubs, deciduous, with alternate, simple, toothed leaves, and many have distinctive, peeling bark. Several species are cultivated for use as ornamentals or for various articles made from the wood. Oil of wintergreen is extracted from the twigs of the sweet birch, *Betula lenta*. One of the best-known species is the canoe, or paper, birch, *Betula papyrifera*. Birches have male and female flowers borne in catkins on separate trees. These flowers serve the purpose of reproduction but do not add anything to the trees' ornamental value. The

14 ❦ *Betula*

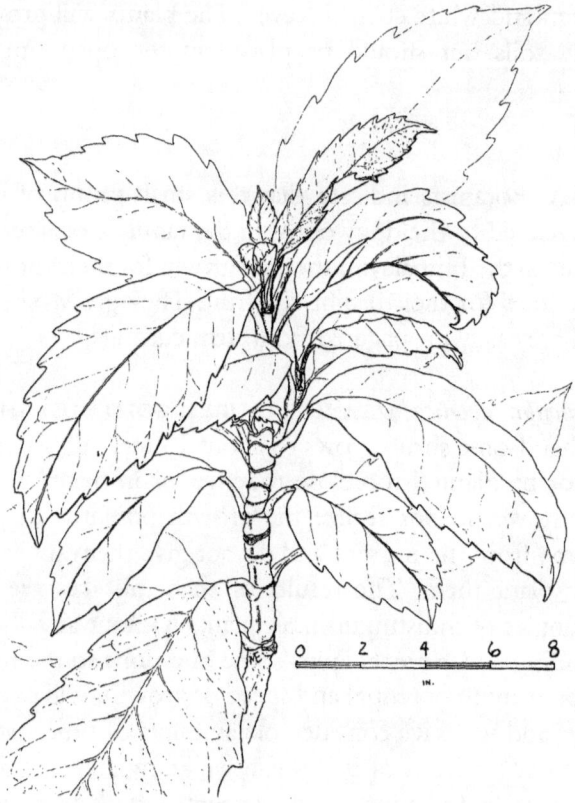

Figure 9. *Aucuba japonica* (Japanese Laurel)

plants may be propagated from seed or from cuttings. Many species of birch are found in nurseries.

———*Betula nigra*, RIVER BIRCH, BLACK BIRCH, RED BIRCH (fig. 10). This native species is the one most commonly found in our area, its distribution ranging from Massachusetts south to Florida and west to Kansas. In nature, it is found in bottomlands along rivers, but it may be cultivated in upland areas as well. The bark peels in papery flakes, mostly silvery on mature stems and reddish brown on younger ones, appearing torn and ragged. The trees may grow to 100 feet but usually don't reach this height. The river birch

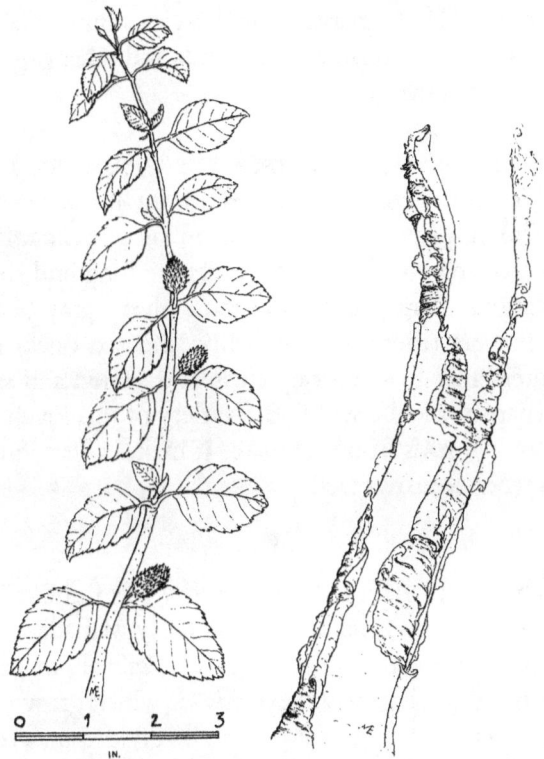

Figure 10. Betula nigra (River Birch)

is now being cultivated more frequently and is used as a specimen or accent tree in sunny to semishaded areas. At planting time, add considerable compost, sphagnum moss, and/or cow manure. Birches respond to 8–8–8 mineral fertilizers added in spring and summer.

❊

BUDDLEJA. Plants in this genus are generally found in more northerly regions, but many people bring their favorite plants with them when they move, which probably accounts for the plants we have seen in our survey of the local region. There are over 100 species in this group, and all of them are called by the common name "butterfly bush." These plants grow best in sunny to semi-shaded locations in soils with considerable humus and other organic matter, such as well-rotted manure.

———*Buddleja davidii,* BUTTERFLY BUSH, SUMMER LILAC. This species, which commemorates the French missionary Pere David, a collector who first found the plants growing in China, is a deciduous shrub growing to 15 feet. Its leaves are long and slender with toothed margins, dark green above and silvery gray beneath. The light blue flowers (there are also white and red ones) have small orange centers and grow in long clusters that nod and sway in the breeze. Perhaps someone will find a variety of this lovely plant that will be more vigorous in our climate. It makes a very fine addition to gardens when it grows well.

❦

BUTIA. This genus name, though it may seem unfamiliar, designates a group of palms, one of which is fairly well adapted to Zone 8. It has been incorrectly called *Cocos*. The Palmae, or palm family, has a large number of genera, very few of which grow outside the tropics. The members of the family can be divided into two groups based on the shape of the leaves or fronds: the feather- or pinnate-leaved palms and the fan-leaved palms. The species discussed here, *Butia capitata,* is one of the feather-leaved types.

———*Butia capitata* PINTO (PINDO) PALM, JELLY PALM (fig. 11). These plants are perhaps better known by the common name jelly palm. They are fairly well adapted to our climate, but some of our winters are just a bit too cold for them. Either the fronds are killed, or the whole plant dies outright. Many representatives of this species have a sufficiently well-protected growing point that they are able to come back after their fronds die from frost. Given this warning, it is well worth trying this palm as a "specimen" plant, where it will stand out from the surrounding plantings, or in rows along a driveway. The graceful, gray-green fronds grow about 10 feet long. The small flowers (appearing in June) are borne in huge clusters and mature into large numbers of bright, yellow-orange fleshy fruits,

Figure 11. Butia capitata (Pindo Palm). Plant 9 ft. tall.

each fruit about one inch across. The flesh is sweet tasting and is sometimes made into jelly, hence the common name.

❀

BUXUS. When recalling the beautiful, old gardens of Virginia, we usually think of well-trimmed boxwood hedges in formal settings. Boxwoods are members of this small genus of evergreens, species of which exude a pleasant musky odor in summer. Unfortunately, the species we raise here does not exude nearly so much of this pleasant odor.

——————*Buxus microphylla* var. *japonica,* JAPANESE BOXWOOD. Hedges grown with this species and variety of boxwood are almost as satisfactory as hedges of *Buxus sempervirens* (common boxwood), which is the species most likely to be encountered in Virginia. However, *B. microphylla* var. *japonica* does not grow as tall as the other species, and the plants are more compact. The plants are very susceptible to nematodes in sandy soils. Otherwise, the hedges are still very handsome, even though slow-growing. The plants are adaptable to a variety of soils and can be grown in partial shade. One may substitute dwarf varieties of yaupon (*Ilex vomitoria*) and achieve as good or better results in our area.

❀

CALLICARPA. This is a group of about 125 different species of shrubs in the verbena family. They all share the rank odor common to the family. Gardeners may not have considered cultivating these plants because they are so common and seem to grow just about everywhere. But they are beautiful (or at least our plant is) and have the added attraction that they provide abundant food for birds. They are native to the Deep South.

————*Callicarpa americana,* BEAUTYBERRY, AMERICAN MULBERRY, FRENCH MULBERRY (fig. 12). The beautyberry has leaves about 3 to 5 inches long, oval or elliptic, with sawtooth edges. The small flowers are borne in clusters in the axils of the leaves, usually in large numbers on each bush. The inconspicuous flowers are followed by clusters of very showy, small, round fruits, light to dark purplish-red. These fruits stay on the plant late into the winter, even after the leaves are shed, not only brightening the landscape but providing welcome food for birds at times when little else seems to be available. These plants sprout as volunteers everywhere, and if you want a plant somewhere else, just dig it up and move it. They are very rugged. After the fruits are gone, you can prune the plants severely and they will reward you with much more vigorous shoots, leaves, and berries next season.

❀

CALLISTEMON. These plants, originating in Australia, are quite sensitive to cold but come back readily from the roots after a hard

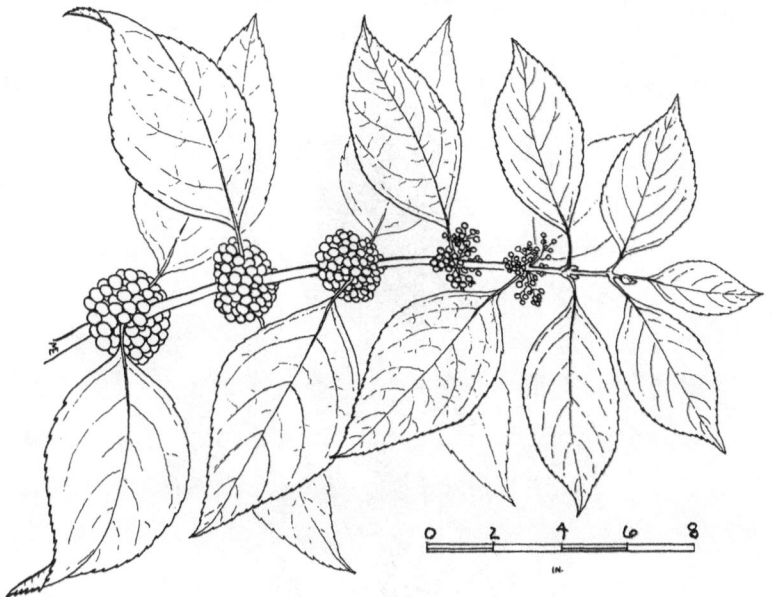

Figure 12. Callicarpa americana (Beautyberry)

freeze, particularly if they are grown in protected places and mulched heavily in the fall and winter. They can be grown from hardwood cuttings or from seed collected in bags enclosing the fruits. There are several species in this genus but the following is most commonly found in the Deep South.

————*Callistemon citrinus* or *C. viminalis*, CRIMSON BOTTLEBRUSH (fig. 13). The two species are very similar. As the common name implies, the clusters of red flowers give the appearance of bottlebrushes, and these make showy displays at the end of the many stems. The plants are evergreen with strap-shaped leaves. The flowers attract hummingbirds and butterflies.

❦

CALYCANTHUS. There are one or two species of these native North American deciduous shrubs. The species listed below can be found in the woods in most of Zone 8 and in many home yards.

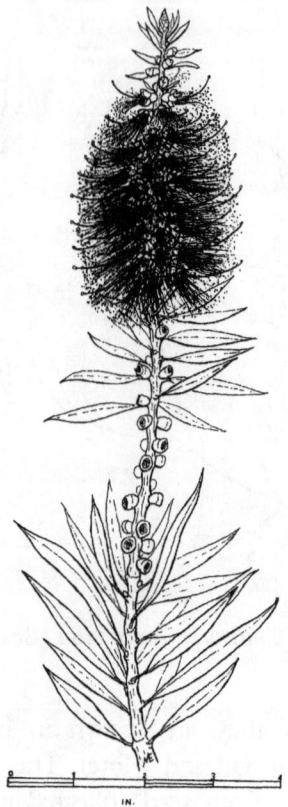

Figure 13. Callistemon citrinus (Crimson Bottlebrush)

Plants have a spreading habit, sending out runners that put up shoots all over the place. The plants require plenty of room and must be cut back often or else they will cover a large area. They are worth the effort!

————*Calycanthus floridus*, SWEET SHRUB (fig. 14). The dark red flowers, each about 1 inch across, have many strap-shaped petals. Their sweet aroma may be strong or weak, depending on individual variation. The fruits are interesting and look like elongated, brown paper balloons, about an inch across with many seeds inside each "balloon."

Figure 14. Calycanthus floridus (Sweet Shrub)

❀

CAMELLIA. Interestingly, the flowering evergreen shrubs of this great group are known by their scientific names. The genus *Camellia* has long been cultivated, not only the two major ornamental species considered native here (although they actually come from eastern Asia) but also a species of the same genus that yields one of the world's greatest beverages, tea. The glossy evergreen leaves, broadly elliptic to ovate, provide handsome accents to our gardens even when the plants are not in flower. One of the great joys of these plants is that they flower at times when most other plants are thinking about closing up for the winter. They start flowering in August, flower more abundantly in September, October, and November, and many are still in full flower through January. (Of course, one must cultivate several varieties of camellias to have blooms throughout this long flowering time.) In Zone 8, the majority of the plants will withstand freezes if they are not too severe. All but the hardiest will succumb to hard, long periods of below-freezing temperatures. But these hard freezes only whet our appetite and we look forward to the next year when the weather, we hope, will return to "normal."

As with most familiar and loved garden plants, there are societies or clubs devoted to the study of the development of new varieties of camellias. These organizations can be found in nearly every cosmopolitan area, particularly in the Southeast and the states of Washington, Oregon, and California, where these shrubs grow luxuriantly. These organizations have encouraged research. One of the most interesting recent developments is the practice of "gibbing," where a plant hormone, gibberellin, is placed on the buds of selected plants. As a result of the hormonal treatment, much larger flowers are produced, even larger ones when all but the treated buds on the same stem are removed. Propagation is usually practiced by taking cuttings of the current season's growth in summer and grafting them on suitable rootstock. Camellias will grow in a variety of conditions, from open sun to heavy shade, but they grow best under pine trees, which provide just the right amount of light shade. The plants seem to be relatively free of destructive diseases and bugs that chew on them, but, like all living organisms, they do have their share.

Camellias are such magnificent plants that they have been illustrated and described in thousands of publications. Some of the best botanical and horticultural artists have made exquisite color illustrations. For this reason the various types of camellias are very well known, thereby obviating the need for illustrations of the group in this reference.

―――――*Camellia japonica* does not have any one common name, but literally thousands of cultivated varieties have been developed over the years. This eastern Asiatic species (Japanese or Chinese) now grows in temperate regions with considerable moisture in all suitable climatic zones around the world, both the northern and southern hemispheres. The shrubs, which may grow to 15 or 20 feet tall, are usually pruned to lower heights to increase their flowering vigor. They have an attractive, light gray, smooth bark, and their usual growth form is rather globular. The plants can be used as hedges and continue to flower under severe pruning. The flowers in nearly wild plants are provided with a single row of petals, but in selected varieties they have become doubled with decreased numbers of bright yellow stamens. Petals may be white, pink, or red with many variations in shades and tones and frequent striations, giving a large range of attractive flowers that may reach 3 to 4 inches in diameter, even without the application of gibberellin. Flowering starts in some varieties as early as July, but usually in August and September. The majority bloom in November, December, and January, tapering off rapidly in February and March. Though relatively few seeds are produced in the advanced varieties, seeds do occur with sufficient frequency to make hybridization possible.

There are countless varieties of camellias, some very well known, others familiar only to fanciers. Many public gardens display camellias and people who have not had the opportunity to see these wonderful plants in their various forms would do well to visit a public garden before selecting a variety for their personal gardens. Also, local growers are usually most willing to share information on camellia varieties and proper culture.

———*Camellia sasanqua*. While there are fewer varieties of the sasanqua, these are equally familiar to gardeners in Zone 8. Sasanqua resembles *Camellia japonica* in general appearance but is characteristically smaller with smaller leaves, finer twigs, and slightly smaller flowers, although such generalizations about plants are often very likely to be inaccurate. The flowers may be white or pink, in varying shades, but without a true red. Sasanquas tend to flower earlier than the japonicas, preponderantly in September and October, when they fill in a gap in flowering time. The same cultural practices apply for sasanqua as for japonica. Most gardens would do well to have at least one, or a few, of these lovely plants.

CAMPSIS. This vine is another native of the Southeast. The trumpet vine grows in profusion both in the open woods and in gardens, sometimes invited, sometimes just by chance. The plants need no special care, but be prepared to work hard to keep them from spreading. This vine has tenacious roots that will send up new shoots time after time.

―――*Campsis radicans,* TRUMPET CREEPER, TRUMPET VINE (fig. 15). Everywhere from eastern North America west to eastern Texas and northward, this vigorous vine appears by roadsides, climbing up telephone poles and any other erect object. Starting in June (perhaps even earlier in Zone 8b) and continuing through the summer,

Figure 15. Campsis radicans (Trumpet Creeper)

the orange-red, trumpet-shaped flowers, 1–2 inches long, are produced in clusters. If you need a vine in your landscape scheme, and desire this color of flower and heed the warning about spreading, you will certainly have no trouble with this beautiful plant. Another value of these plants is their attractiveness to hummingbirds. It has been recently reported that the trumpet vine has poisonous leaves.

❧

CARYA. These are the hickories familiar to most Southerners. They are very tall trees bearing nuts with very hard shells. We might not know that that beloved nut, the pecan, is in the same genus. Other than the fact that its nut has a rather thin shell, the pecan resembles other hickories in the appearance of its fruits, leaves, and bark, and the way the flowers develop. There are maybe eight or ten species that can be found in our region, some of them rare, limited in their distribution, and found in habitats not suited to human habitation.

Two of the three species included here are natives over much of the eastern United States (and over much of the Deep South). The third species, the pecan, is native to the Mississippi River drainage, from Illinois south to Texas and thence into Mexico, but is not native to the eastern part of Zone 8. Pecans, now seemingly wild in the eastern areas, were brought in from the west and north, probably by Native Americans in prehistoric times. Further spread of the trees was aided by squirrels. Seedlings mostly have rather small nuts, and only cultivated selections have the large, fleshy nuts that are easy to crack. In most states where pecans are grown, county agents can provide literature listing the best-suited varieties and hints on cultural practices. If you wish to grow one of the hickories, either start plants from seeds (layered in sand over winter, then planted out) or get a plant from a nursery.

Though it may be tempting to dig up a small tree, aspiring transplanters will think they're headed for China before they get to the bottom of the taproot!

———*Carya glabra,* PIGNUT HICKORY (fig. 16). This species grows over much of the eastern United States, usually in hardwood forests or mixed hardwood and softwood forests. It has no apparent pref-

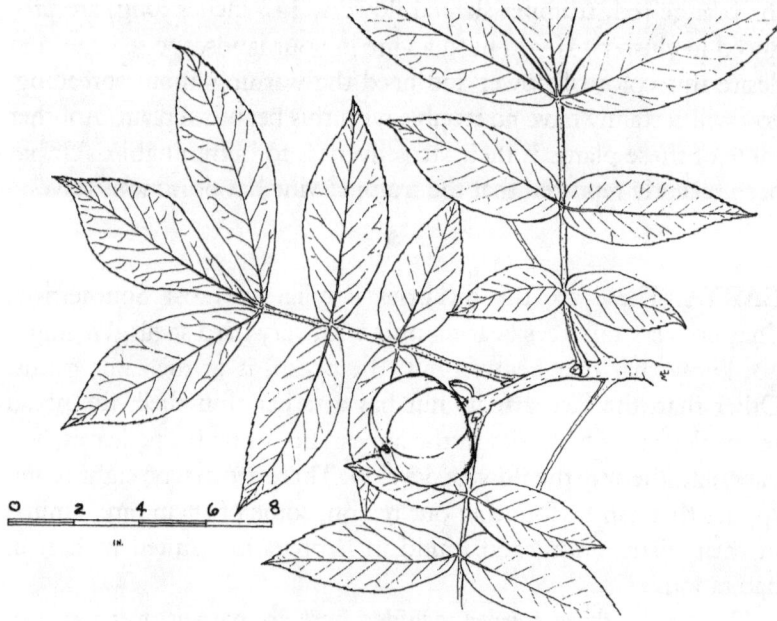

Figure 16. Carya glabra (Pignut Hickory)

erence for any specific soil type. It is a very tall tree, and one of the forest dominants. "Pignut" seems to refer to the usually small nuts, but there is one variety (*C. glabra megacarpa*) with larger fruits. In the fall, this species has very striking, bright-yellow foliage. The leaf is usually divided into 5–7 leaflets that are narrower than the other species listed here. It has relatively smooth bark. One hazard associated with this hickory as well as others is that rotary gas- or electric-powered mowers can pick up the fallen nuts and shoot them out like bullets.

———*Carya illinoinensis,* PECAN (fig. 17). We are almost certain that someone will insist that we have misspelled the species name, but this is the spelling given by the original authority, which must be followed. As mentioned above, pecans were brought to our region from farther west and north where this species is a native of the Mississippi River Valley. The green fruits look very much like other hickories and, when mature, dry and split open, revealing the

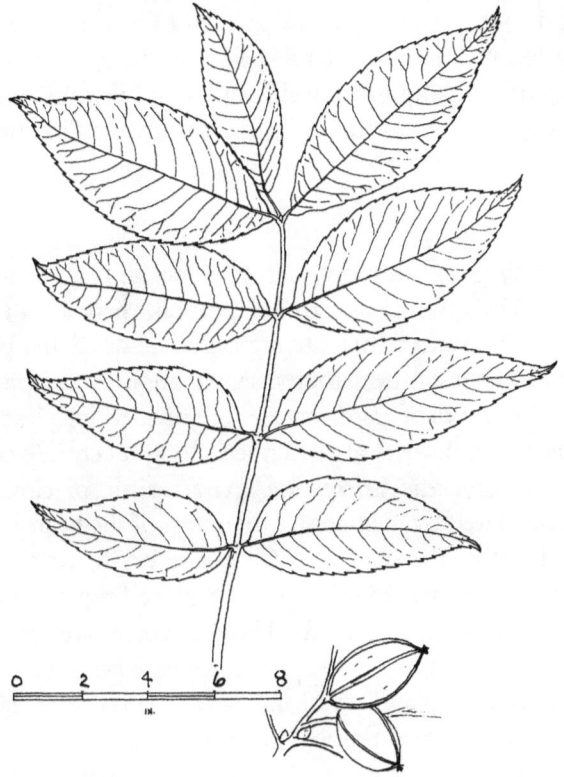

Figure 17. Carya illinoinensis (Pecan)

brown nuts within. Pecan lumber has a good grain pattern for furniture.

Pecans are rather messy trees with male catkins, produced in great numbers, shedding after the pollen has been produced. If the trees are too close to the house, these catkins will fall on the roof, get into eaves troughs and stop up the drains. In the fall when the leaves shed, individual leaflets fall separately, leaving a petiole and midrib of the leaf as a long, flexible, slender object that is guaranteed to try the patience of those trying to get rid of them.

———*Carya tomentosa,* MOCKERNUT HICKORY. The common name is said to come from the very hard, thick-walled nuts that deter the efforts of squirrels to get into the meat. This is another widespread species in woodlands. It is not known so much for its fall coloration but, like all the hickories, produces very good wood for various uses, including shovel handles and various other implements that require wood that is tough but gives a little under pressure.

❦

CASSIA. This genus of more than 500 species is a member of the bean family. Though mostly a tropical group, there are a few representatives in the warm temperate regions. The dried and pulverized leaves of one of these latter species, senna, are used as a purgative in medicine. The leaves of *Cassia* species are divided into leaflets, from few to many. The leaflets are arranged along a central stalk, like a feather. They have the interesting characteristic of closing up at night, a trait shared by many other members of the bean family. The flowers of most of the species of *Cassia* are bright yellow. The following two species are found in Zone 8 more frequently along the coast—Zone 8b—than inland. Though there are undoubtedly common names for these two species, those whom we have asked shrug their shoulders and say "just cassia." So try your hand at inventing an interesting common name for them!

———*Cassia corymbosa* (fig. 18). Shrubs of this species have rather slender stems and grow to about 8 feet tall. The leaves are evergreen, with 2–4 pairs of leaflets; the leaflets are linear, or strap shaped. The leaflet shape distinguishes this species from *C. coluteoides,* listed next. Another telling characteristic is the shape of the fruit—a dry capsule, or legume—which is about 5 inches long and round in cross section. The flowers are rather similar to those of the next species and are bright yellow. Cultivation is easy, requiring little more effort than good watering while the plant is getting established. This species grows rather well in soil to which quite a bit of humus has been added.

———*Cassia coluteoides* (fig. 19). This cassia is also a shrub, 5–15 feet tall, with a little more robust stem than the former species. It, too, is an evergreen, but a rather tender one. A sunny location on

Figure 18. Cassia corymbosa (Cassia)

the south side of a windbreak will give it a better chance to survive severe freezes. This species' sensitivity to frost probably accounts for the fact that it is less seen inland than near the coast. The leaflets of this cassia are rounded, in three to five pairs. The fruit is a dry legume, segmented, with one seed per segment. The pod is flattened, in contrast to that of *C. corymbosa,* above. This species is undemanding and, after establishment, will take all kinds of punishment.

CASTANEA. The well-known chestnut blight has essentially wiped out the American chestnut. The Chinese chestnut is a good tree, but the nuts aren't up to the same standard as the American. There is one other species of *Castanea,* a native, that we seldom

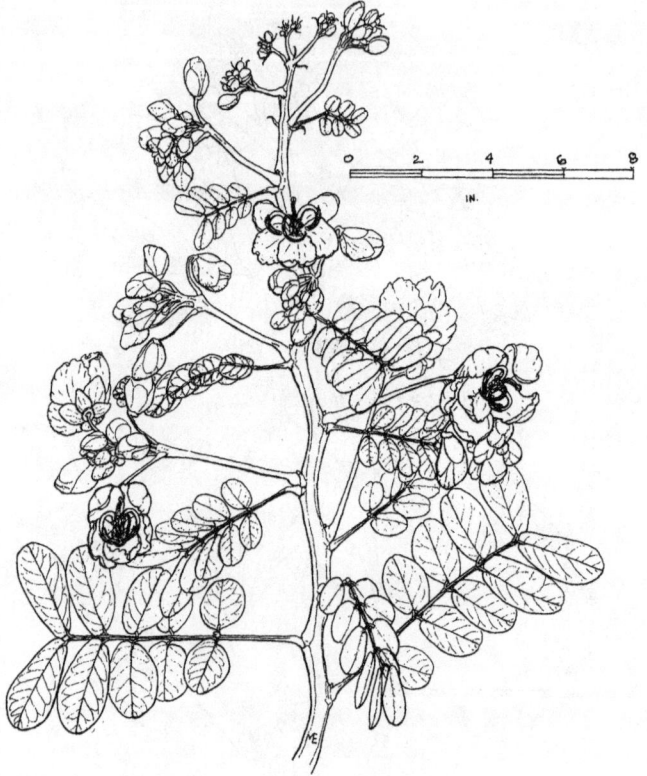

Figure 19. Cassia coluteoides (Cassia)

associate with the chestnut, and that is the chinquapin. All species in *Castanea* are rather handsome deciduous trees, and the two mentioned here are very much at home in our climate. They are included in this work for their value as shade trees, but the Chinese species also produces a satisfactory hardwood for lumber. The small individual flowers occur in fingerlike clusters near the tips of branches. Flowers of both species produce a disagreeable, sickeningly sweet aroma. Cultivation is quite simple; early care and watering until the plants are established is about all that's required.

———*Castanea mollissima,* CHINESE CHESTNUT (fig. 20 a and b). This species grows about 60 feet tall and has a habit similar to the

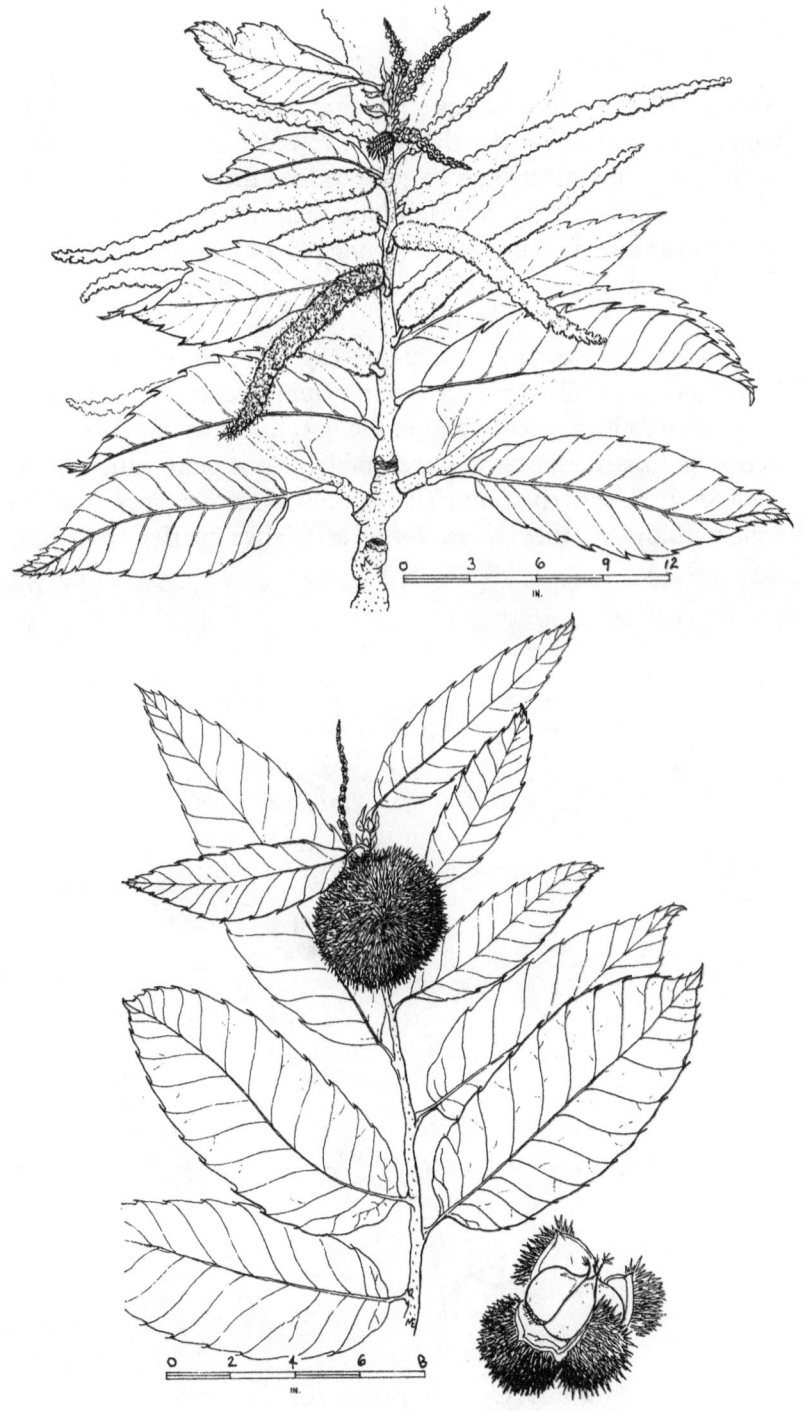

Figures 20 a and b. Castanea mollissima (Chinese Chestnut)
20a–Flower. 20b–Fruit

American species. It produces a nut that is not at all comparable to the American chestnut but that some consider edible. One person familiar with the plants says he used to feed the nuts to hogs. The leaves are elliptic, about 6 inches long, with rather large serrations on the margins. We have seen only one plant in our trips, but there must be many more that have escaped our attention.

———*Castanea pumila*, CHINQUAPIN (fig. 21). One of the authors, D. J. Rogers, recalls climbing a chinquapin tree as a boy, opening the prickly fruit, then chewing on the raw, half-inch nut. His recollection is that the flavor was acceptable, if not quite equal to the favored American chestnut. This species does not appear to be widely cultivated these days, however. Native to the Southeast,

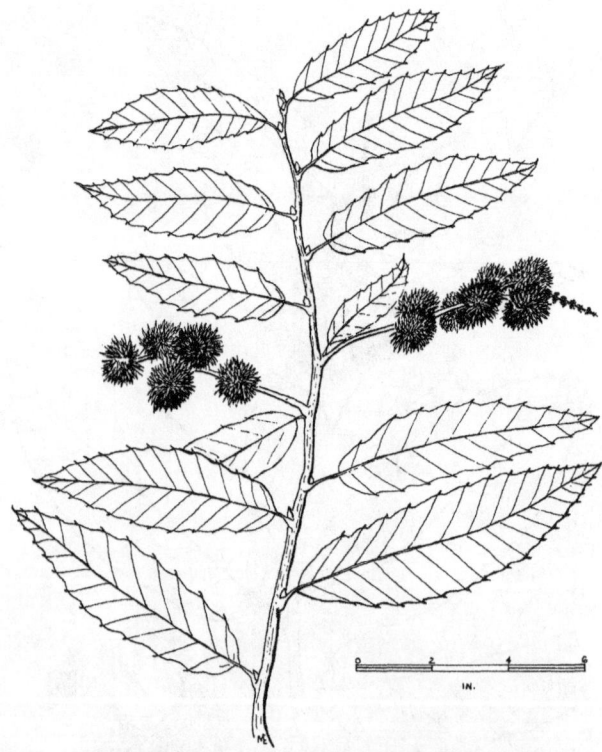

Figure 21. Castanea pumila (Chinquapin)

these trees are rather small, semi-globular in outline, no more than 20–25 feet tall, and quite frequent in rich woods. The white catkins appear in June or July and stand out from the surrounding vegetation.

❁

CATALPA. Trees of this genus are so well known in the Deep South that they are included here only because, if left out, someone would surely note the omission. They are the "worm trees" so useful for fishermen in the summertime. The leaves are regularly attacked by large caterpillars which, if left alone, will chew up every green thing on the tree. For any other plant, pesticide would be in order, but *never* for the worm tree! Left to its own devices, the tree will grow a new batch of leaves anyway.

———*Catalpa bignonioides,* CATALPA, WORM TREE (fig. 22). This tree, which may grow 50–60 feet tall a little farther north, is found everywhere in the Southeast and may be cultivated all over the country. It has big, heart-shaped leaves borne on long stalks. It is deciduous. The flowers are very attractive, mostly white with colored spots in the tube of the corolla. Planting is unnecessary if mature trees already occur, because the prolific seeds will yield volunteers in no time. The plants grow rapidly.

❁

CEDRUS. This genus of conifers has three or four species in it, all of which can be used in horticulture, but only one is grown in the Deep South. The Cedar of Lebanon (*Cedrus libanensis*) is perhaps one of the most famous for its historical value as the tree used to build Solomon's Temple. It would be nice to grow it here, but it tends to want a bit cooler climate than ours, such as Zone 7 (refer to the temperature map and growing zones in the front of the book). Most species in this genus can be grown from cuttings taken from mature young limb-tips and rooted in a good rooting mixture (such as perlite combined with some peat moss). To get good specimens, it is best to buy container-grown plants from your nursery. Transplanting, if it has to be done, is best accomplished in the

Figure 22. Catalpa bignonioides (Catalpa)

spring just before the new growth starts. Try to get as much of the root system, intact, as possible, and keep the transplants well watered during the first year of growth. After this, a handful of 8–8–8 fertilizer scattered around the base of the tree will be sufficient for plants up to 5 feet tall. The plants are not normally susceptible to attacks by insects or by fungi. If they do become infected, contact the county agent (or your handy farm supply store).

———*Cedrus deodara*, DEODAR (fig. 23). The species name, deodara, comes from a Sanskrit word meaning, literally, "tree of the gods." The species is a native of western Nepal and adjacent areas in China. It can grow as tall as 150 feet, and is a very highly prized timber tree in both countries. There are several cultivated varieties

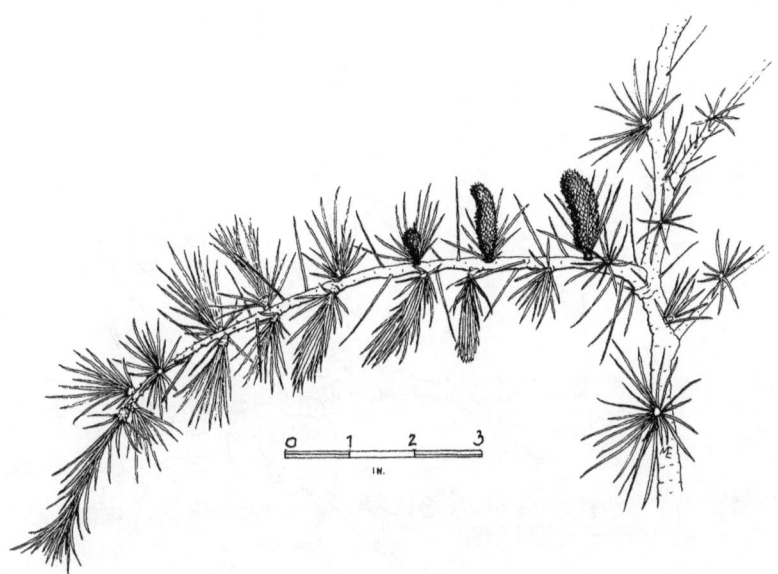

Figure 23. Cedrus deodara (Deodar)

grown in this country for horticultural purposes. The one most commonly encountered in Zone 8 is the cultivar 'Fontinalis' which is probably the same as 'Pendula'. We have seen a specimen more than 40 feet tall, but there is no reason to think this is the tallest this variation will get to be. The needle-leaves are 1½–2 inches long and clothe most of the younger stems with a handsome, gray-green foliage. Several other cultivars are known, some of which are dwarfs.

❀

CERCIS. This is also a member of the bean family and prolifically produces pods with little brown seeds. There aren't many species in this group, but they are found as natives in North America, Europe, and the Middle East.

─────*Cercis canadensis*, REDBUD (fig. 24). One of the first trees to flower in spring, the redbud, native to Zone 8, is found in rich woodlands. It is easily spotted when blooming because it produces hundreds of small, pink flowers along its stems. It is a beautiful, graceful, small tree with arching branches and very handsome heart-

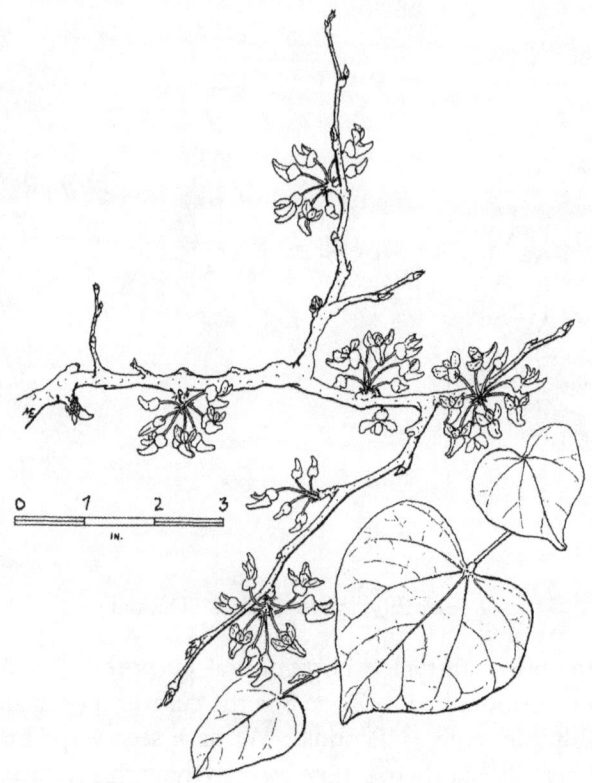

Figure 24. Cercis canadensis (Redbud)

shaped leaves. The plants grow all up and down the eastern side of North America, as far west as eastern Texas and northward. Several cultivars have been introduced, some with more intense color, others double-flowering. Very little extra care is needed for these plants, but they respond to twice-a-year application of a balanced fertilizer, either 6–6–6 or 8–8–8, plus regular watering during dry spells.

❀

CHAENOMELES. This genus of three species belongs to the rose family (Rosaceae), and its members are very closely related to the regular quince, *Cydonia oblonga,* a much larger plant, raised for its fruits. *Chaenomeles* species usually have smaller fruit and are raised only for their flowers.

———*Chaenomeles speciosa*, FLOWERING QUINCE, JAPANESE QUINCE (fig. 25). This species is deciduous with leaves that are glossy-green, have serrate edges, and are usually oval shaped. It may be propagated from seed, from cuttings of the roots, or by layering (please see explanation of layering under *Deutzia*). It is one of our earliest spring-flowering shrubs, growing to about 5 feet tall. The plants look a little "stalky" in winter, but their bright orange-pink flowers are such a pleasure to see that this makes them worth growing. They are very hardy plants and can withstand the hard winters that Zone 8 occasionally endures. These plants will grow in the open sun or in light shade and should be welcome in most gardens. Several new cultivated varieties have been developed and are for sale

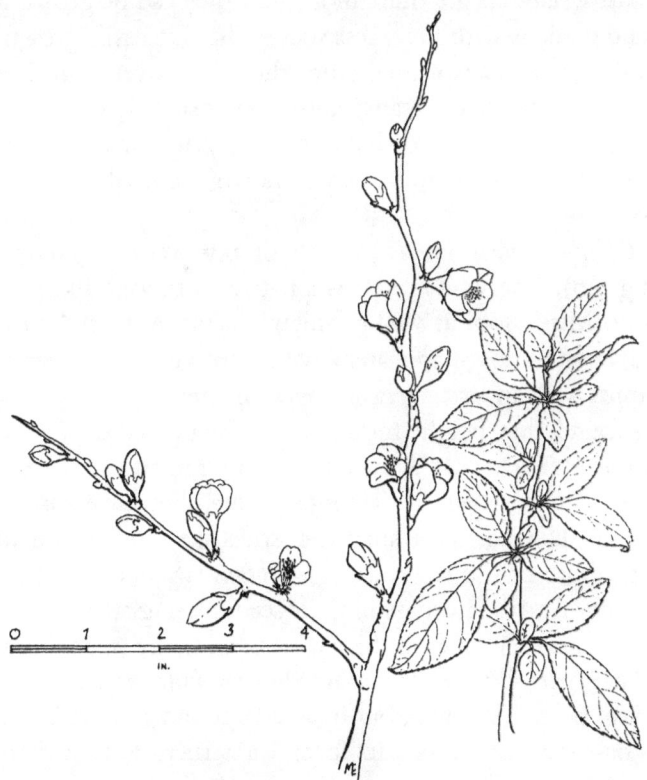

Figure 25. Chaenomeles speciosa (Flowering Quince)

at nurseries. They mostly differ in flower color; for example, 'Jet Trail' has a snow-white flower, 'Cameo' a peach-colored blossom, and 'Texas Scarlet' a good red.

❦

CHAMAECYPARIS. This genus of coniferous trees and shrubs contains only 7 or 8 species. It is related to junipers and to arborvitae. Two of its species are well known in the United States—one in the Northwest, the other in the East, from Florida and Mississippi north to Maine. Both are used as commercial timber but have very nice ornamentals that have been selected and developed. Like many other evergreen conifers, their leaves are normally green but frequently may be variegated yellow-green or white and green. Their native habitats are quite moist, but they can be grown in our yards and gardens without excess water. They are easily grown from seeds or may be transplanted from the wild as very small plants. Care is necessary during transplanting to ensure that the roots are not severed or allowed to dry out. Very few nurseries carry the plants, but they deserve to be better known and used.

————*Chamaecyparis thyoides*. WHITE CEDAR, ATLANTIC WHITE CEDAR (fig. 26). The white cedar is a native from west Florida westward into Mississippi as well as suitable areas far to the north. Its native habitat is low-lying areas with "brown water," but it will grow quite well in drier areas. It is a tall tree, 80–90 feet, but is seldom found that height today because it is used as a source of cedar lumber. The trunks grow tall and straight, and at maturity the trees have a high crown. As younger plants, they have a pyramidal appearance. The leaves are small and scale-like, similar in many respects to those of the red cedars (*Juniperus virginiana*). The white cedars differ from red cedars in their seed-bearing structures: small (¼-inch diameter) cones in the white cedar, fleshy structures in the red cedars. A small number of named ornamental varieties of white cedar have been developed, but by and large the greatest use of this species has been for lumber. Its form, leaf structure, bark characteristics, and general vigor make it a very good possibility for an accent

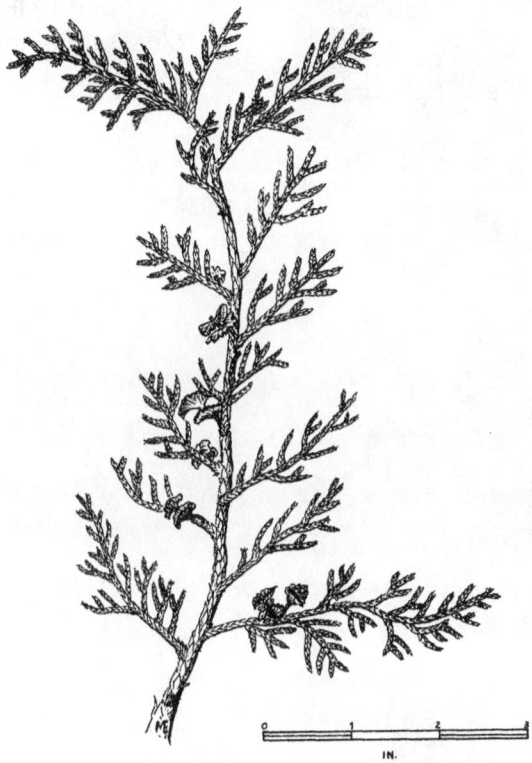

Figure 26. Chamaecyparis thyoides (White Cedar)

tree. Finding one in a nursery is the biggest problem. Though inquiries were made to the Florida Division of Forestry, they do not grow seedlings of the white cedar.

———*Chamaecyparis thyoides* cv. Ericoides (fig. 27 a and b). This striking, cultivated variety of the white cedar has been a problem for botanists and horticulturists as well as for many people who have them in their yards. The plant has been mistakenly placed in the wrong genus (as for example, in Watkins and Sheehan (see selected Reading list) who place it in *Cupressus* and call it Italian cypress). People who raise the plants find that the trees develop unsightly brown, dead branches, and may eventually die. In these cases, there

Figures 27a and b. Chamaecyparis thyoides cv. Ericoides.
27a–Plant, 10 ft. tall.
27b–Detail of branch

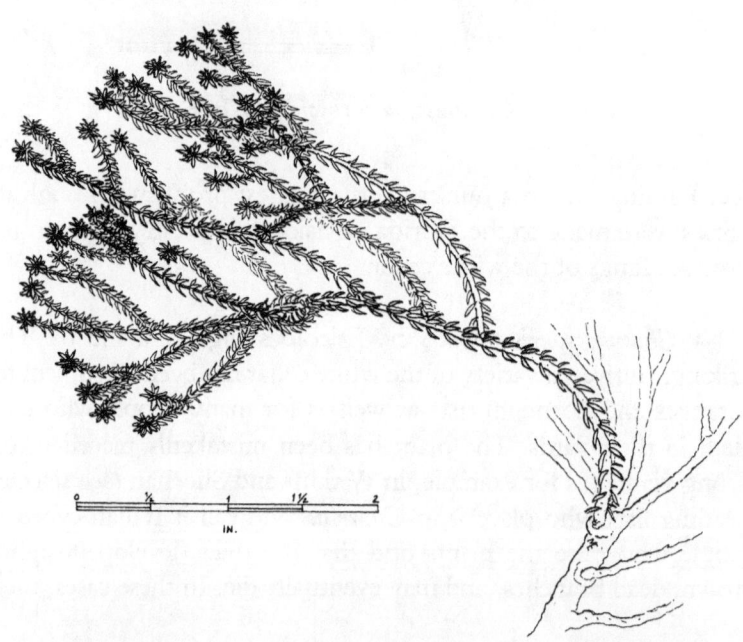

seems to be no evidence of a disease, and it is currently thought that extremes of cold weather may cause the symptoms. Whatever the problem, the trees grow to about 40 feet in a very striking columnar form. The plant needs no special care and grows quite well in sandy soils. A sunny location is advised, on the south side of the house or of some protecting wall.

CHAMAEROPS. This genus of only one species belongs to the palm family. It is native to the Mediterranean region in Europe and North Africa. It is not a very obvious ornamental in Zone 8, possibly because it is a low-growing type, with rather small, palmate (fan-shaped) leaves, and can be easily overlooked.

———*Chamaerops humilis,* EUROPEAN FAN PALM (fig. 28 a and b). This is one of the hardier palms, growing well in open sun. The spines along the petioles of the leaves help to distinguish this palm from others in our region. The soils of its native region are usually much richer loams than our sandier types, so augmentation with plenty of humus, dried manure, and sphagnum moss, or other available combinations will help the plants along. Fertilization with mineral fertilizers (especially nitrogenous and potash types) are important in our soils, but if these are not available, then an 8–8–8 type will do. Plants do not get more than 8 feet tall but will grow into clusters, joined by decumbent stems, that are not too different from the saw palmetto.

CHIMONANTHUS. There are only four species in this genus of shrubs, all imported from China. They grow mostly in warm temperate regions, and the one below does well in the Deep South.

———*Chimonanthus praecox,* WINTERSWEET. This shrub, growing to 10 feet tall, is closely related to our sweet shrub. Its flowers are yellow, fragrant, and, as the common name implies, sometimes occur in midwinter before the leaves appear. The leaves, which may be 6 inches long, are smooth-edged and bright green. This shrub is not

Figures 28a and b. Chamaerops humilis (European Fan Palm)
28a–Plant 8 ft. tall. 28b–Leaf with spines

commonly found in Zone 8, but should be used much more than it is.

CHIONANTHUS. Although the name of this plant is very similar to the one just described (*Chimonanthus,* family Calycanthaceae), it is in a very different family, Oleaceae, the same family as the olive. The difference is best seen when the plants are in fruit. The fruits of the olive and the graybeard are very much alike: an oval, fleshy fruit with a single, stony seed, whereas those of *Chimonanthus* are dry, bag-like fruits with many small seeds. The species of *Chionanthus* described here is a very valuable shrub or small tree, native in our woods.

———*Chionanthus virginicus,* GRANDSIR (GRANCY) GRAYBEARD, FRINGE TREE (fig. 29). These are either small trees or large shrubs.

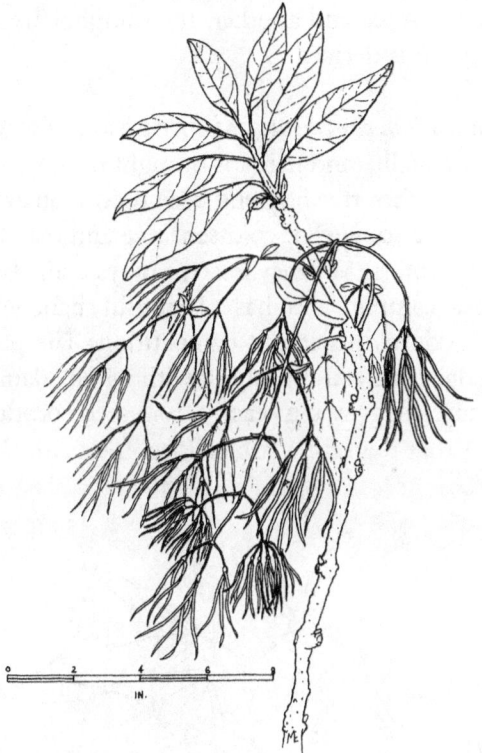

Figure 29. Chionanthus virginicus (Grancy Graybeard)

They have many delicate, white flowers in very large clusters, appearing just after the new green leaves in April and May. Though these plants grow well in either shade or sun, they do best when grown in light shade because then the stems do not become so elongated. Pruning the plants just after flowering will allow more blooms to occur in closer clusters and give a better effect.

❀

CINNAMOMUM. As one can readily see, this genus has a name very similar to one of our favorite spices, cinna*mon*. However, there are other plants besides cinnamon in the same genus, one of which is the well-known camphor tree. The spice plant is not raised here. Most of these plants are of Asiatic origin, including several species that produce the spice and another, the camphor tree, from which the drug camphor is derived.

———*Cinnamomum camphora,* CAMPHOR TREE (fig. 30). Although once used medicinally (and perhaps brought to this country for that purpose) the camphor tree has long been used as an evergreen shade tree. The trees are somewhat cold-sensitive and may be killed back in very severe winters, but they nearly always come back quickly in the spring. The camphor tree has a beautiful shape when given the opportunity to develop without competition. The plants may also be used as hedges and stand pruning without any damage. They are frequent volunteers, growing readily from seeds produced in fleshy, purplish fruits that migrating robins love. If unsure about the iden-

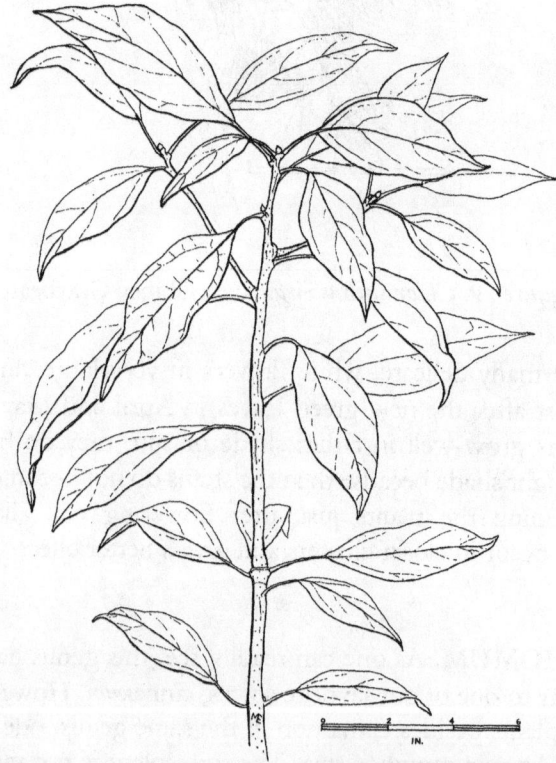

Figure 30. Cinnamomum camphora (Camphor Tree)

tity of a tree suspected of being a camphor, just crush a few of the leaves. Camphor leaves are aromatic and the odor is immediately recognizable.

❊

CITRUS. This is the genus of oranges, grapefruits, lemons, and others. Many of the species have been in cultivation for a long time, and the group is thought to have originated somewhere in southeast Asia. From that region Arab traders introduced citrus (probably several times) to the Arabian Peninsula and the shores of the Mediterranean Sea. The Spanish explorers brought oranges with them when they first came to the New World, and either intentionally or unintentionally planted seeds in many places. Because of a complex genetic mechanism, the seeds will not yield offspring true to type but produce plants that are "throwbacks" to more primitive types. One such throwback that prospered in south Florida was the sour orange, which thrived and produced many plants in localities along the west coast of Florida. Later, when improved fruit was introduced, growers found that the sour orange made a better rootstock than did the roots of the improved varieties, so nearly all of our citrus is grafted on sour orange rootstock. Citrus groves have been planted in many areas of Zone 8, and many types thrived only to be killed back by extreme, freezing weather. Such weather occurs rarely, but often enough to discourage continuation of cultivation. One type that has thrived better than others in Zone 8 is the mandarin orange, described below.

————*Citrus reticulata*, MANDARIN, SATSUMA. Several other common names are used for what is best known as the mandarin orange, but satsuma is the name more familiar to many. The species is a little more cold-tolerant than the sweet orange, but still requires considerable effort to keep the almost certain freezes from killing it. The rind or skin of the satsuma pulls away from the segments of the flesh much more easily than that of the sweet orange. Also, the segments separate readily, making it a delightful fruit to eat out-of-hand. Because of a disastrous disease of citrus, citrus canker, plants raised in Florida cannot be shipped out of state. In fact, buyers must sign certificates when they purchase plants, certifying that the plant will be planted in Florida and not taken out of the state. Similarly, satsumas—or any other citrus for that matter—may not be shipped into Florida. In spite of these difficulties, the dark green, evergreen foliage, the lovely white flowers, and the bright orange fruits make these excellent decorative plants. They are worth all the effort, and may provide delicious fruits as well.

CLYTOSTOMA. This is a small genus of mostly tropical, evergreen shrubs or vines from South America, in the family Bignoniaceae. Several well-known trees and vines cultivated in yards and gardens are members of this family. One of the most common trees in the Deep South, catalpa, is well-loved by fishermen who get their worms for bait off its leaves. Another member of the family Bignoniaceae, a vine, is *Campsis radicans,* the trumpet creeper or trumpet vine. The flowers of the genus *Clytostoma* are about the same size as the trumpet vine flowers.

————*Clytostoma callistegioides,* ARGENTINE TRUMPET VINE, LOVE CHARM (fig. 31). This species is an evergreen vine with opposite leaves; each leaf is elliptic-oblong, about 4 inches long. The plants climb by tendrils. The handsome, tubular flowers are light purple or mauve with white throats, flowering freely in early summer. The plants should be raised in the sun, on a fence or arbor, preferably on the south side of a wind-break or wall since they are quite tender. Otherwise, its cultivation is similar to that of most other plants—it requires watering in dry spells, and fertilizing once or twice a year with a standard, moderate formula such as 8–8–8. The plants may be propagated by cuttings.

CORNUS. This is a genus of the dogwoods, and there are a number of handsome plants in this genus in addition to the beautiful white and pink species known so well. Plants in this genus vary in stature from trees 40–50 feet tall to shrubs and perennials growing just at ground level, such as the Canadian dogwood. *Cornus* species

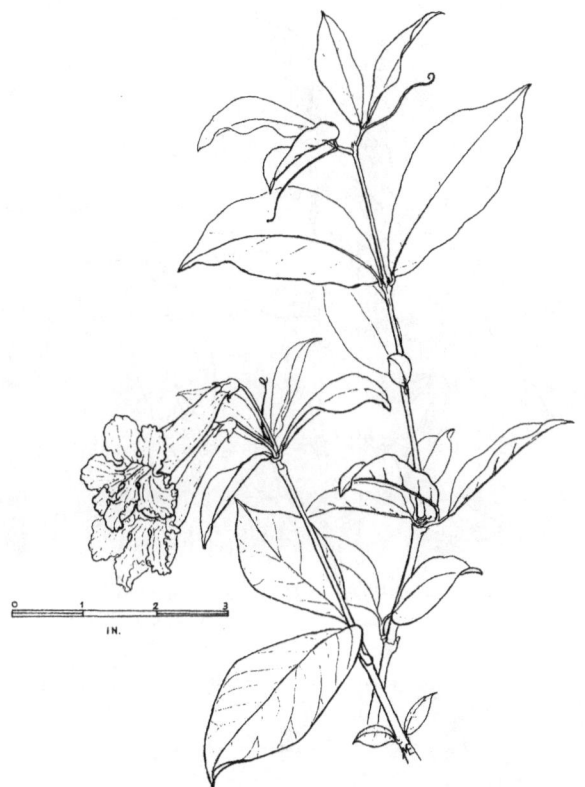

Figure 31. Clytostoma callistegioides (Argentine Trumpet Vine)

occur both in North America and in Asia. Dogwoods may be propagated from seeds, by cuttings from mature wood, by layering (see procedure for layering under *Deutzia*) or by bud-grafting. While they are not too demanding in their cultivation, they are particularly sensitive to drought and must be watered regularly during their early years.

————*Cornus florida,* FLOWERING DOGWOOD, DOGWOOD (fig. 32). People are often confused when told that the showy parts they associate with flowers are not parts of the flower itself, but are actually *bracts,* or modified leaves. If they examine the "flower" of the dogwood carefully, they will find a tight cluster of numerous, small,

Figure 32. Cornus florida (Dogwood)

inconspicuous flowers surrounded by white bracts. No matter, these conspicuous structures are still beautiful and provide a great show in early springtime. Only white-flowering dogwoods are native to Zone 8, but farther to the north, both pink- and white-flowering ones are found. Fortunately, however, both will grow very well in our Deep South gardens. There are several varieties. A scale insect attacks some plants, shortening their lives. A Volk oil spray in late winter or early spring will help keep the scale down.

COTINUS. There is only one species in this genus, the one mentioned below. This genus belongs to the same family as poison ivy and mango but does not have any rash-causing tendencies. We raise this shrub for its very attractive, large flower clusters. Though the individual flowers are tiny, collectively they are both handsome and interesting. These plants grow in our sandy soils without any undue problems. They do respond to at least twice-yearly applications of mineral or organic fertilizers and regular watering. They may be propagated by seeds, root cuttings, or layering (please see description of layering under *Deutzia*).

————*Cotinus coggygria,* SMOKE TREE (fig. 33). This attractive shrub or small tree does not deserve such a jaw-breaking species name. "Smoke tree" is a good descriptive term for the visual effect of the plant in bloom and in fruit. The arrangement of countless small flowers in each cluster creates this effect, the flowers' collective colors giving the impression of smoke. The whole large shrub will

Figure 33. Cotinus coggygria (Smoke Tree)

have numbers of these globular-shaped inflorescences (flower clusters) distributed from top to bottom of the plant (which may be up to 15 feet tall). Plants flower in May. Though this species is infrequently planted in Zone 8, it should have a larger place in the "bank" of interesting and attractive shrubs.

❦

CRATAEGUS. This is a very interesting group of shrubs or small trees, most thorn-bearing, with a rather large number of named species. This interest, in addition to their horticultural value, is that they develop fruit without the usual combination of egg and sperm. This process is known to botanists as apomixis. Apomixis produces a number of variant offspring that may or may not retain the characteristics of the parents. Until this process was discovered, botanists named the new variants as new species. We really don't know how many native species actually exist. Several may have horticultural merit, but we have seen only one native, the following, actually grown in gardens. It is recognizable and not too variable.

———*Crataegus uniflora*, HAWTHORN, THORN, THORN APPLE, HAW (fig. 34). One of the attractive features of this deciduous shrub or small tree is the characteristic twisting or bending of the stem, giving it the appearance of a Japanese bonsai. Another good feature is that it will grow in the driest, deepest sand, a characteristic more and more valuable in these days of overpopulation and increasing demand for water. These plants don't appear very often in gardens though it seems they are "tolerated" in areas that have been cleared of brush. The branches from the main stem produce a "zig-zig" pattern and are well provided with short, sharp spines. Although the flowers are borne singly in the axils of the leaves, each plant produces a large number of small flowers, ½-¾ inch wide, with white petals. These usually occur in the early spring shortly after the redbud and red maple have bloomed. Unless the plants are very small, they are difficult to transplant because they have a very long tap root. The fruits are rather small, reddish-greenish, but do not make much of a show.

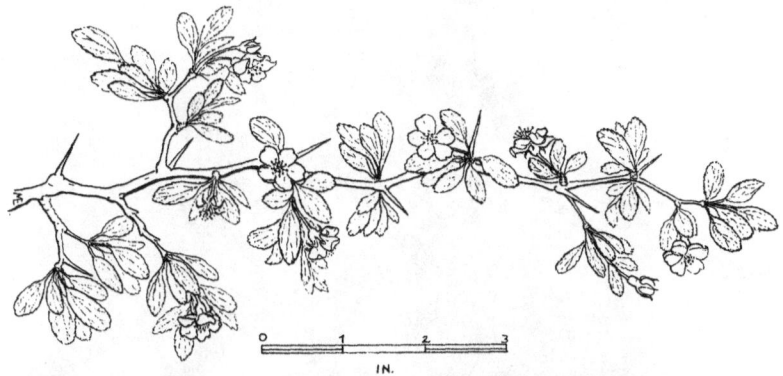

Figure 34. Crataegus uniflora (Hawthorn)

❦

CRYPTOMERIA. This genus, of only one evergreen species, belongs to the group of plants known as gymnosperms. Among the gymnosperms are the pines, bald cypresses, spruces, firs, yews, and several others, all sharing the characteristic that their seeds are essentially exposed to the open and produced, in most species, in cones of varying shapes and sizes. The one species of *Cryptomeria* is grown for timber in Japan and used extensively in horticulture.

―――*Cryptomeria japonica,* JAPANESE CEDAR (fig. 35). Several different species of gymnosperms are commonly called "cedar," indicating the difficulty one has using common terms to differentiate groups of plants. For example, the red cedar of Zone 8 is a member of the genus *Juniperus* and is quite different from *Cryptomeria japonica.* The Japanese cedar grows to a height of 150 feet. The tree is pyramidal in outline, very handsome if grown in the open and not crowded. The evergreen leaves are awl-shaped, about ½-inch long, arranged spirally around the stem. The leaves grow close together giving the stem the appearance of a narrow cylinder. The seed-bearing cones are globose, about 1 inch in diameter. The tip of each scale of the cone has 2 or 3 curved appendages. Because of the eventual size of this species, one should give it plenty of room to grow,

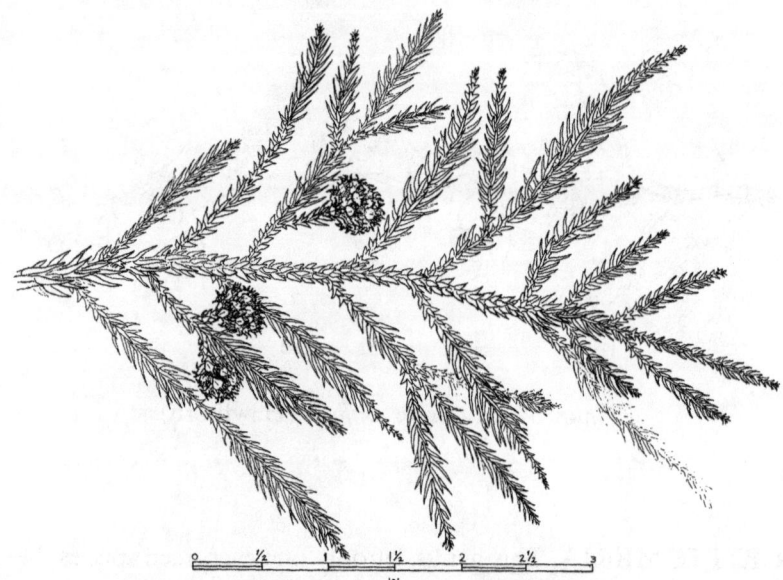

Figure 35. Cryptomeria japonica (Japanese cedar)

using it as a very handsome accent or specimen plant. The plants are not sensitive to cold and grow equally well in Zone 7, just to the north of our zone. Normal soil preparation for planting and regular applications of nitrogen and potassium, particularly, are recommended. Several cultivated varieties are known.

CUNNINGHAMIA. There are two species of this genus of evergreen, coniferous (cone-bearing) trees from China and Taiwan in cultivation. They are somewhat related to the bald cypress in southern swamps but, unlike this species, the cunninghamias are evergreen. There are quite a few of these trees planted in Zone 8, even though they are marginally hardy. Perhaps the appeal of their glossy foliage and symmetry is so great that people are willing to take the chance that their trees will escape the freezes. So far as we know, it appears that these plants do not reproduce very well in Zone 8, so are generally purchased from nurseries.

———*Cunninghamia lanceolata,* CHINA FIR, MONKEY TREE (fig. 36). These trees will, under ideal growing conditions, reach heights of 100 feet or more. We seldom have any taller than 40 or 50 feet, and most are under twenty, because of the coldkill. The evergreen leaves are awl-shaped, pointed, and spiral around the stem in a very attractive arrangement. We have a tendency to plant these fine trees too close to other plantings not realizing their potential for growth. These trees should definitely be placed in a well-sheltered, sunny location, and protected against both cold and wind as much as possible.

❦

CYCAS. Although of a palm-like habit, this genus is not really related to true palms but belongs to an entirely different group of plants. The cycads are an "evolutionary hangover"; that is, they were a much more dominant part of the earth's vegetation during the epochs when coal-forming layers were being produced, 300 million years ago.

Figure 36 a and b. Cunninghamia lanceolata (China Fir)
36a–Plant 60 ft. tall. 36b–Cones 3 in. long

———*Cycas revoluta,* SAGO PALM, KING SAGO (fig. 37). This species has evergreen, frond-like leaves at the top of a rather stubby trunk, and generally does not grow more than 8–10 feet tall, the crown of leaves borne on a trunk up to 1 foot in diameter. The leaves are about 3 feet long and are stiff and shiny dark green. They have been used as background fronds in large floral arrangements, as well as in the artistic Japanese arrangements known as bonsai. The word "sago" refers to a starchy food substance that can be derived from the trunks of the plants. The plants are somewhat sensitive to the very low freezing temperatures that we have occasionally, but will usually come back after the freeze. These plants should be used as "stand-alone" accents, in a corner of the garden or yard where they will not interfere with, nor be screened out by, other plantings. One may hear people refer to "king" sago palm. This designation probably derives from the fact that the male cones (see fig. 37) are quite striking, but only occur occasionally.

❦

DEUTZIA. There are about 40 species of shrubs in this genus, most from Asia and many in cultivation. One species is used in the

Figure 37. Cycas revoluta (Sago Palm). Plant 3 ft., cone 2.5 ft.

Deep South. It usually flowers in April and perhaps in May, along with or a little after the mock oranges. These shrubs, like many others, should only be pruned after they have flowered so that the plants will have a chance to put on new growth, on which much of the flowering occurs. They can be grown in the sun or in partial shade, but full sun will probably give the most profuse flowering. The plants are best propagated from cuttings of either green or hard wood. They may also be layered (bend down a healthy branch, cover it with soil and leave the branch attached to the parent until roots have formed along the portion underground).

———*Deutzia scabra,* cv. Candidissima (fig. 38). There is no known common name for these plants and they are all generally

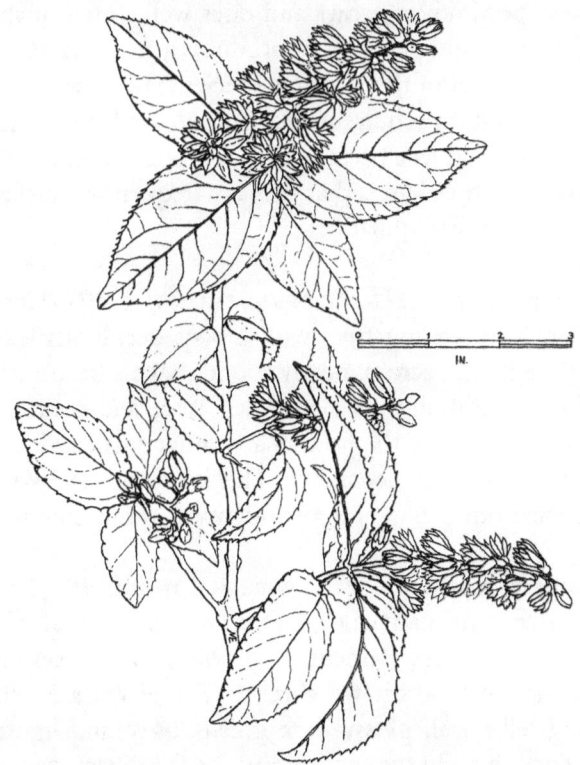

Figure 38. Deutzia scabra cv. Candidissima (Deutzia)

referred to as "deutzia," which is pronounced either as "doitzia" or "dootsia". Plants left alone or neglected for a number of years may grow to be 15 feet tall. They become rather gangly at such heights and produce fewer flowers. The double-flowered variety mentioned here has some light pink in the outer petals, but flowers are mostly white. The flowers grow in crowded racemes along the tips of side branches, making a handsome show.

❀

DIOSPYROS. This is an interesting group of trees, more tropical than temperate, found in North America as well as Africa, Europe, and Asia. Ebony, the well-known black or dark brown wood used in carvings, bowls, etc., is found in Africa. In Asia, or, more specifically, China and Japan, as well as in California and the Deep South, the Japanese persimmon grows and does well. In all of the southeastern U.S., the wild persimmon produces a fruit that is edible but not as choice as that of the Japanese species. These two species have no special cultivation requirements. Indeed, we have trouble keeping just one fruiting tree because they reproduce and grow so well. About their only problem is that they are frequently attacked by the web-worm or tent caterpillar.

———*Diospyros kaki,* JAPANESE PERSIMMON (fig. 39). This is a medium to small, spreading tree, with glossy, deciduous leaves. The fruits are 2–3 inches across, nearly round, have a beautiful, irridescent, reddish-yellowish-pink skin when ripe, and contain two or three large seeds. While the flowers are rather small and not showy, the fruits are really very decorative. This is one of the favorite fruits of the Japanese but is only infrequently used in this country.

———*Diospyros virginiana,* WILD PERSIMMON (fig. 40). This is such a common tree in the native flora of the Deep South that it is seldom considered a decorative element in our yards, but it certainly is as handsome as some others we cherish. The leaves are about 5–6 inches long, elliptical, glossy dark green above and lighter green beneath. They shine in the sunlight. In the fall the leaves turn either bright yellow or dark red and add considerable color to our land-

Elaeagnus 🌸 57

Figure 39. Diospyros kaki (Japanese Persimmon)

scape. The fruits of the wild persimmon are quite tasty when fully ripe, but will really pucker your mouth if only slightly green. Fruits have an orange skin and usually occur in large numbers on a tree. Possums and coons really love them.

🌸

ELAEAGNUS. This genus is composed of a few cultivated species and some 40 others that occur naturally in southern Europe and Asia, with one species in North America. Almost all are recognizable by the silvery color of the leaves, which have star-shaped scales on the upper and lower surfaces. These plants thrive in our sandy soils and will withstand drought quite well. They are evergreen. The flowers are generally small and sweet-scented.

———*Elaeagnus pungens* cv. Simonii, SILVER THORN, THORNY ELAEAGNUS (fig. 41). The silver thorn is a tall shrub used mostly in hedges, for which it is particularly well suited. They grow quite densely and form an almost impenetrable barrier, augmented by

58 ❧ *Eriobotrya*

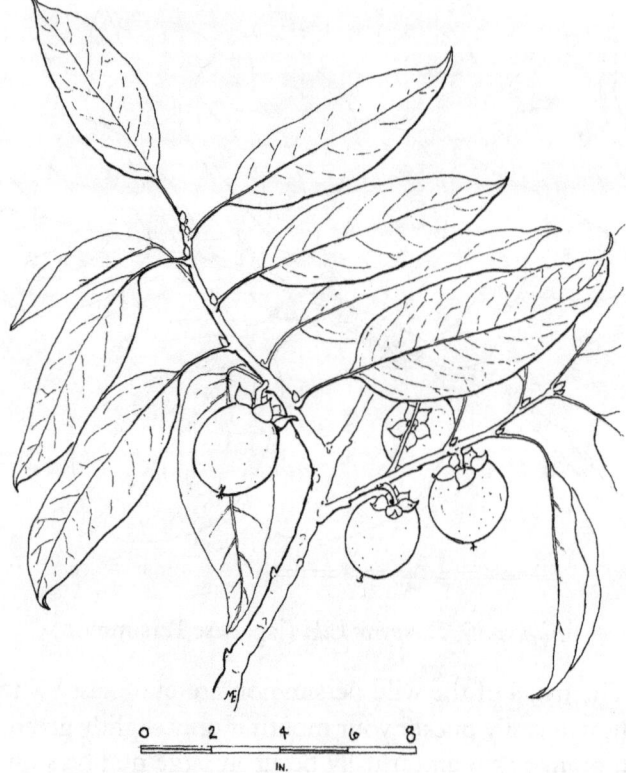

Figure 40. Diospyros virginiana (Wild Persimmon)

their sharp thorns on the stems. The flowers are scented though rather inconspicuous; these are followed by ellipsoid fruits about 1 inch long. The plants can be grown in full sun or partial shade, and thrive in both conditions. The Florida Department of Transportation has planted the silver thorn in the median of Interstate 10, near Pensacola. The plants are used in many other settings, either neatly pruned or allowed to grow naturally.

ERIOBOTRYA. This small group belongs to the same family as the roses, for technical reasons. The species are all originally from eastern Asia and Japan, a fact that emphasizes again the interna-

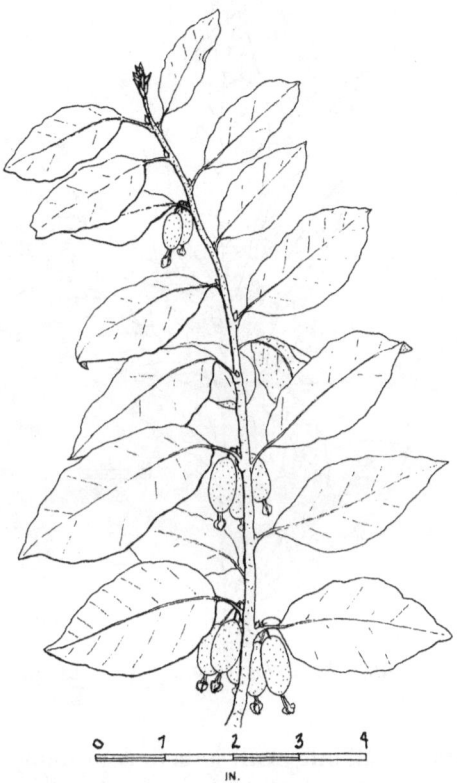

Figure 41. Elaeagnus pungens cv. Simonii (Silver Thorn)

tional flavor of Deep South gardens. We have one species of *Eriobotrya*, the following, of importance to our gardens and yards. The plants can be propagated either from seed or from cuttings, the former being the easier propagation.

———*Eriobotrya japonica*, LOQUAT, JAPANESE PLUM (fig. 42). This large shrub or small tree has been used in Southern gardens for quite a long time. It not only provides very handsome foliage, but also produces numbers of rather acid fruits. The very fragrant flowers are rather insignificant, but the yellow fruits, which usually have fuzzy hairs on them, are quite decorative. The fruit makes good jellies and pie. The severe cold of Zone 8 winters in the late 1980s

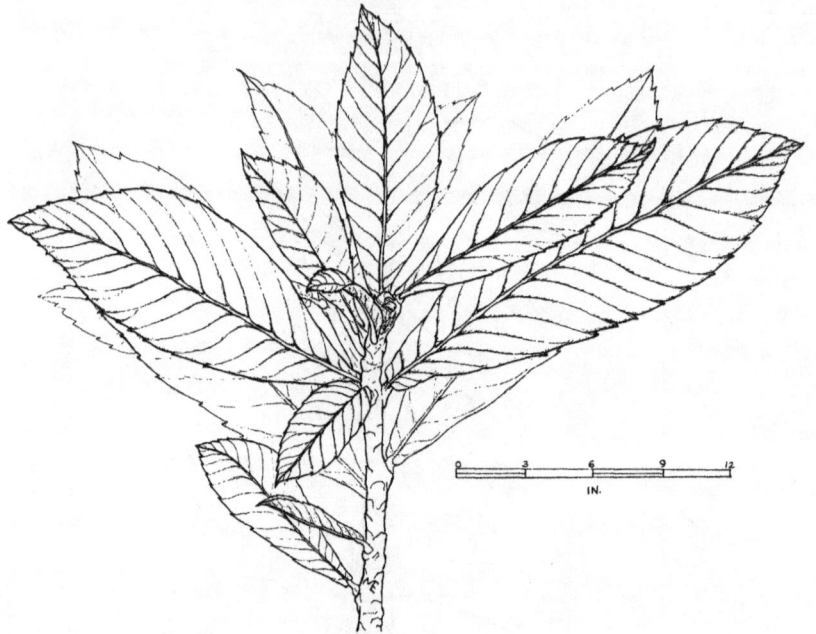

Figure 42. Eriobotrya japonica (Loquat)

caused some dieback but, with judicious pruning, the trees can be kept in an acceptable shape. Also, frosts alone may sometimes prevent flowers from setting fruit. The loquat is very susceptible to fire blight, but this can be prevented by using the appropriate fungicidal spray. Consult a local gardening center for the best fungicide to use and the timing of applications.

❦

EUCALYPTUS. This very large genus (more than 500 species) is in the family Myrtaceae, a group that includes trees and shrubs mainly of Australia. There, they are the dominant tree; the tallest of them challenge the California redwoods as the tallest trees in the world. More than 200 species have been introduced into other regions around the world and have become so well adapted that they almost seem to be a part of the native flora. In the U.S., most have been used in California, where they make popular roadside plantings and produce striking and beautiful flowers. Only one species can be grown in our climate, the one listed below. Unfortunately, it is quite sensitive to cold weather and is often killed back to the ground.

————*Eucalyptus polyanthemos (?),* SILVER DOLLAR TREE (fig. 43). This small tree gets its name from the interesting fact that the tree produces juvenile leaves that are quite different from the mature leaves. The juvenile leaves are rounded and covered with wax, and they grow opposite each other on the stem, the paired opposite leaves giving the appearance of a completely rounded structure. The latter characteristic, plus the fact that the juvenile leaves are all silver-gray (masking the chlorophyll underneath), is the reason for the

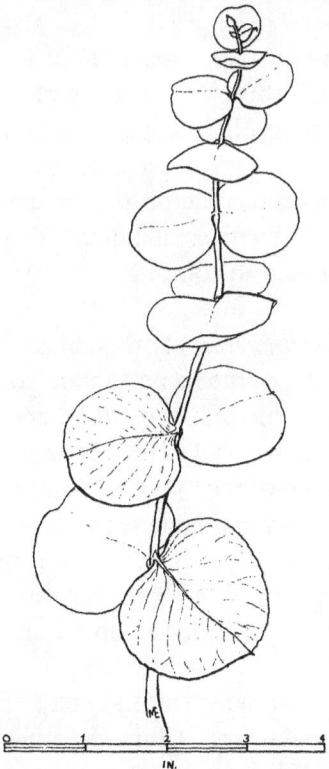

Figure 43. Eucalyptus polyanthemos (?) (Silver Dollar Tree)

common name silver dollar tree. This is such a striking plant that it is worth trying in your garden. Perhaps growing it in a sunny, protected spot would help. The cut stems, with their unusual leaves, make handsome arrangements and are frequently used by florists.

❦

EUONYMUS. (Pronounced "U-ón-i-mus.") In the Deep South we cultivate at least three species of this genus of 170 species (mostly Asiatic), two of which are native Americans. The group includes small trees, shrubs, and creepers. They are used as background planting, ground covers, or hedges. They are either evergreen or deciduous, with oval to elliptic leaves. The plants do not require any special attention other than the usual addition of fertilizer in early spring or late winter and again in the summer.

————*Euonymus americanus,* STRAWBERRY BUSH, BURSTING HEART (fig. 44). This shrubby species is native in eastern North America, reaching the southern edge of its range in north Florida. The flowers are small and inconspicuous, but the fruit at maturity accounts for either of the common names. The outer walls of the fruit are red and covered with small bumps; they split open to show a rounded seed with reddish coloring.

————*Euonymus atropurpureus,* WAHOO, BURNING BUSH. We don't know how this species got the name "wahoo" or what it means, but the second common name tells something about the leaves—they are very bright red and colorful in the fall and are, of course, deciduous. This is the second of the two American species, native in eastern North America, much the same as the previous one. We haven't seen but one or two bushes of this plant, nor can we really say very much about it, except one should try to balance the seasonal color of the garden by using this plant to help out in the fall.

————*Euonymus fortunei,* WINTER CREEPER (with many varieties in cultivation). This species, from China, is either a small shrub or a creeper with evergreen leaves and some roots modified as holdfasts like ivy. According to Watkins and Wolfe (1986)—see "Selected

Euonymus 63

Figure 44. Euonymus americanus (Strawberry Bush)

Reading" list—these plants make good ground covers under trees on the "heavier soils of northwest Florida."

————*Euonymus japonicus,* SPINDLE TREE (fig. 45). Most variations we have seen are not trees, but handsome evergreen shrubs. The leaves are closely set, with no evident open stem space on the branches, ovate or elliptic and several other shapes; the margins are serrate. Variations include all green, green and white, yellow and green (or gold and green) foliage. The plants may be used as basal

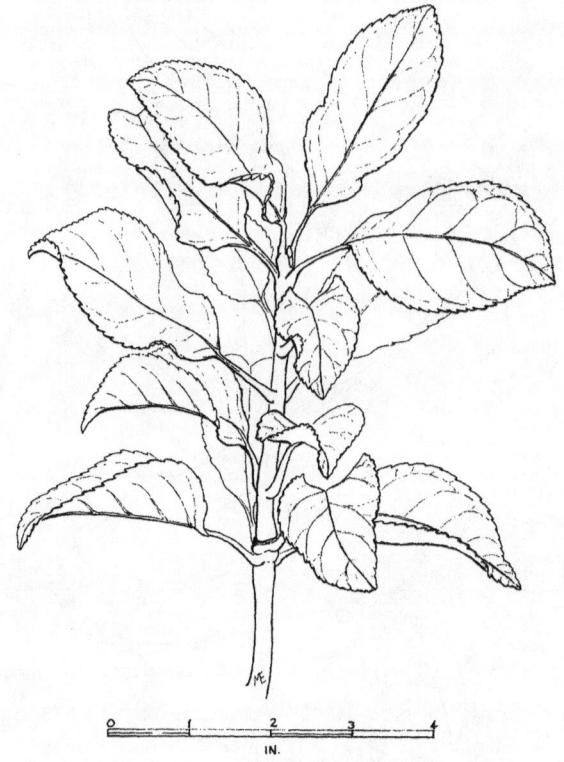

Figure 45. Euonymus japonicus (Spindle Tree)

planting around the house or may provide accent in an otherwise green area. One sees many more of the variegated types than the all green ones.

❦

FATSIA. This is a genus with only one species. The name is said to be derived from a Japanese word. It belongs to the same family, Araliaceae, as the rice paper plant *Tetrapanax*.

———*Fatsia japonica,* JAPANESE FATSIA, FORMOSA RICE TREE, PAPER PLANT (fig. 46). The Japanese fatsia is a shrubby species raised entirely for its attractive, large, glossy leaves. The stems either have no branches or only a few. The leaves are carried on foot-long petioles and are about 1 foot across. They are cut into 5–9 lobes and

Figure 46. Fatsia japonica (Japanese Fatsia)

are glossy green. Though small, white flowers do appear, they are not outstanding and are followed by small black fruits. The plants may be propagated from seeds or from cuttings, preferably in the summer time. Fatsias make good fill-in plants in sheltered, shady areas (they are frost-sensitive) on the south side of a house or wall, perhaps on each side of an entrance way. Some gardeners keep the plants in pots which they take in during the coldest weather, but during the rest of the year leave in the ground.

❦

FEIJOA. There are only two species in this genus, from the southern parts of Brazil, Paraguay and Argentina. They are mostly subtropical, but the one described here is grown extensively in Zone 8.

———— *Feijoa sellowiana*, PINEAPPLE GUAVA, FEIJOA (fig. 47). This is an evergreen shrub often used in hedges or as a background plant but fitting for individual display in a sunny location. The plants will grow to about 18 feet but most often are kept lower by judicious pruning. The flowers are carried along the upper stems, usually several on a branch, and they are white with many bright red stamens. The flower contrasts nicely with the light, evergreen, elliptic to oblong leaves, which grow up to 3 inches long. The white petals are fleshy, or succulent, and are rather sweet tasting. The fruits, which may be made into jelly, are green with some red, 2–3 inches long. This handsome shrub grows well in our soils and seems to be able to weather winter extremes nicely. The plants may be propagated

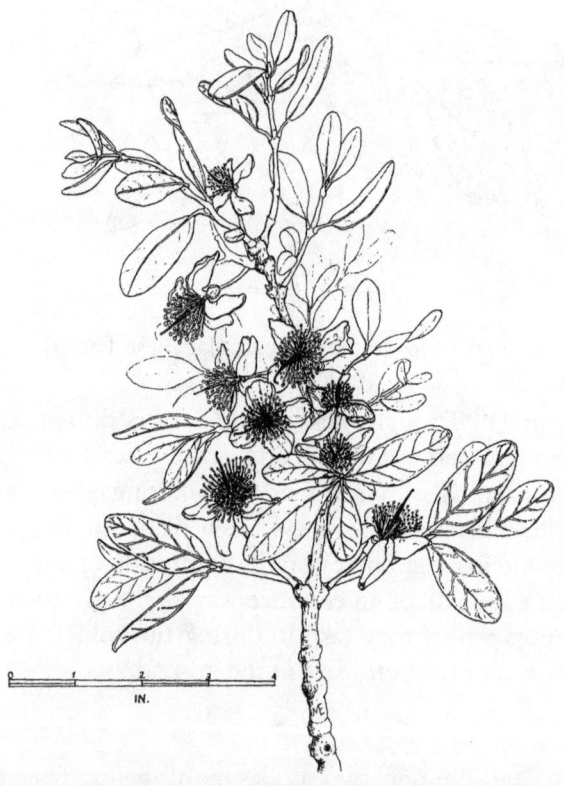

Figure 47. Feijoa sellowiana (Pineapple Guava)

either by seed, air-layering, or by cuttings rooted from young wood. (Air-layering is accomplished by girdling a vigorous young shoot, applying a root growth compound, covering with moist peat moss, then wrapping and tying plastic tightly above and below the girdled area. Roots will form in a month or two.)

❦

FIRMIANA. There are between 10 and 15 species (depending on which authority is used), of this genus in Africa, southeast Asia, and China. The genus is a member of the same family (Sterculiaceae), that chocolate comes from, though the species of this genus do not have fruits resembling those of the chocolate tree. Several of the species can be grown for shade, as street trees, or ornamentals; they propagate easily from seeds.

———*Firmiana simplex,* CHINESE PARASOL TREE, CHINESE BOTTLE TREE, JAPANESE VARNISH TREE, PHOENIX TREE (fig. 48). This is a handsome tree that grows up to 60 feet tall (though the ones we have seen are no more than 30 feet). The bark of the tree is smooth and light green. The leaves are very large, up to 12 inches across, with the petiole (leaf stalk) up to 20 inches long; they are shiny above and lighter green below. There are generally 3 lobes on each leaf. The flowers occur in large clusters and are not particularly attractive. Each flower, about ½-inch across, is light brown. Typically, the two sexes occur in separate flowers in the same cluster. The fruit, which hangs down like a little brown parasol, gives this species one of its common names. Many seeds are produced on each tree, and they germinate readily around the parent tree. For this reason there tends to be a cluster of young plants around the parent tree unless these are mowed down regularly. Although the trees are perfectly hardy in Zone 8 (they have withstood three of the hardest winters in history), we have not encountered very many specimens in this area.

❦

FORSYTHIA. This is a very small genus with no more than six or seven species, all but one of which are Asiatic. The exception is from

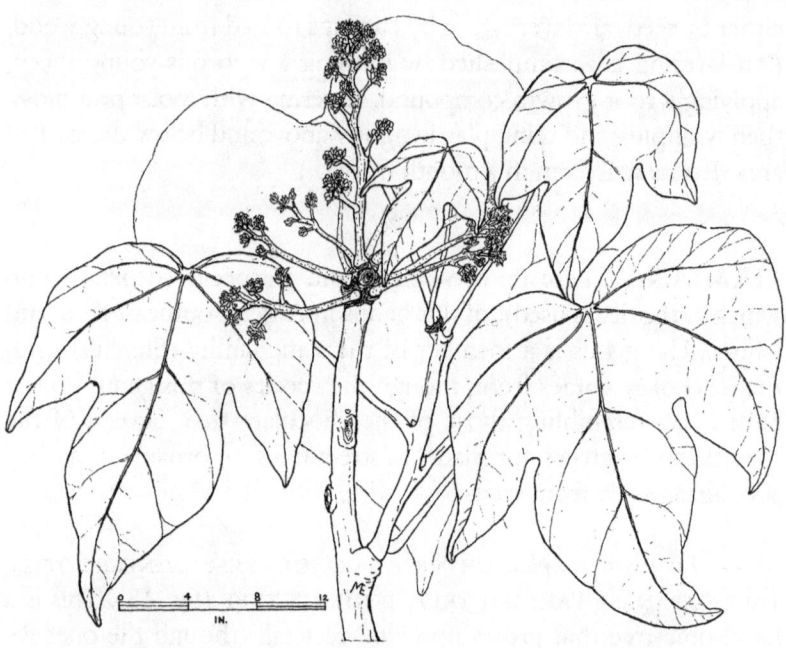

Figure 48. Firmiana simplex (Chinese Parasol Tree)

Albania. These plants are better suited to more northern regions than Zone 8 but, as is frequently the case, they are grown here. Perhaps someday selections will be made so that forsythias in the Deep South will give the same, wonderful burst of early spring flowers as those that occur in the middle states and northward into New England. They are not very demanding plants and grow quite well in the Deep South climate.

———*Forsythia* × *intermedia,* GOLDEN-BELLS (fig. 49). As indicated before, the *x* between the generic and specific terms indicates a hybrid that has become very prominent in our gardens. In this case, two Chinese species were crossed to make a decorative plant that is very striking in the early spring, when masses of golden-yellow flowers cover its arching stems. Unfortunately, in Zone 8, the plants do not seem to bear nearly as many blooms per stem as those grown farther north, and the flowers themselves are not nearly

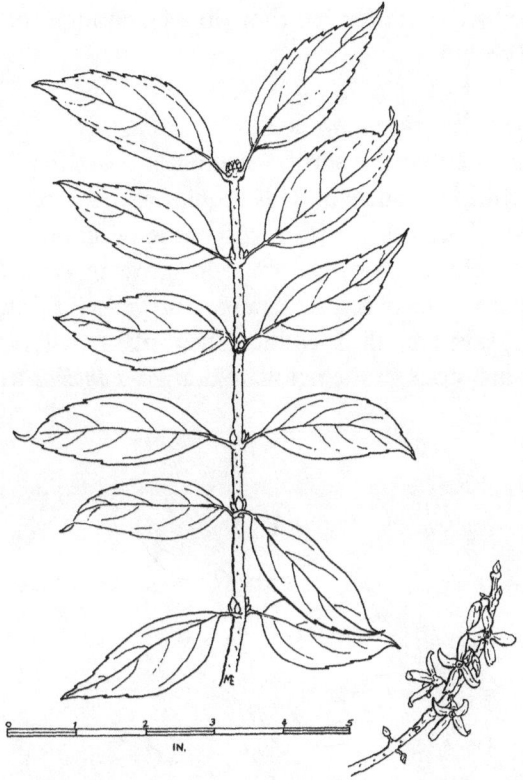

Figure 49. Forsythia × *intermedia* (Golden Bells)

as vibrant a color. A number of cultivated hybrid variations are known. We are not sure which of these our plants represent.

FORTUNELLA. This small genus, probably native to east Asia and the Malay Peninsula, is close kin to the citrus. There are only a few species known and the exact locale of their origin is unknown because they have been cultivated so long that many places look as if they might be the native habitat. Members of this genus are evergreen with attractive, glossy green leaves. They have flowers and fruits much like all other citrus. We have two species in Zone 8, and though they are quite tender, they do well enough that fairly large shrubs are grown.

———*Fortunella japonica,* ROUND KUMQUAT (fig. 50). This kumquat is an evergreen shrub usually not more than 6–8 feet tall and quite thickly branched. This species and the one that follows are a bit more hardy than their relatives the sweet orange and grapefruit, which accounts for their ability to grow in Zone 8. The fruits may be up to 1 inch across, orange- or yellow-skinned, and the whole fruit, skin and all, is edible. These plants will do better here if grafted onto stock of the trifoliate orange, *Poncirus trifoliata*.

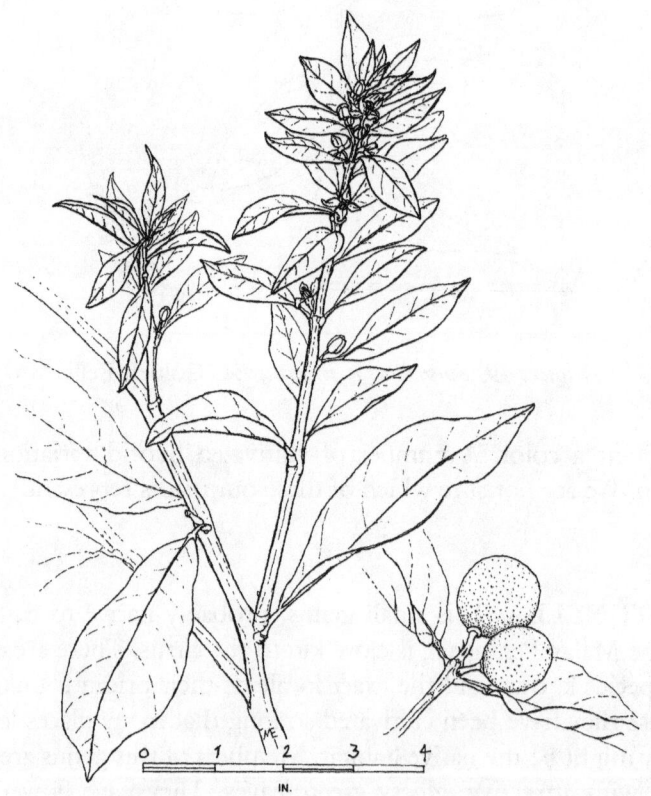

Figure 50. Fortunella japonica (Round Kumquat)

———*Fortunella margarita,* OVAL KUMQUAT. This is also grown in Zone 8 and has many of the same characteristics as the plant described above except that this species grows a little taller than the round kumquat. It also tends to have fewer thorns than the former, and the branches aren't so densely placed (though this is hard to judge at times). Fruits of both species may be eaten raw or made into marmalade, and the plants are attractive ornamentals.

❧

FRANKLINIA. This genus of only one species belongs to Theaceae, the tea family. The genus *Camellia* is also a well-known group in this family.

———*Franklinia alatamaha,* FRANKLIN TREE (fig. 51). This is a very famous species for two reasons: it was discovered by the famous explorer John Bartram and named for Benjamin Franklin; it

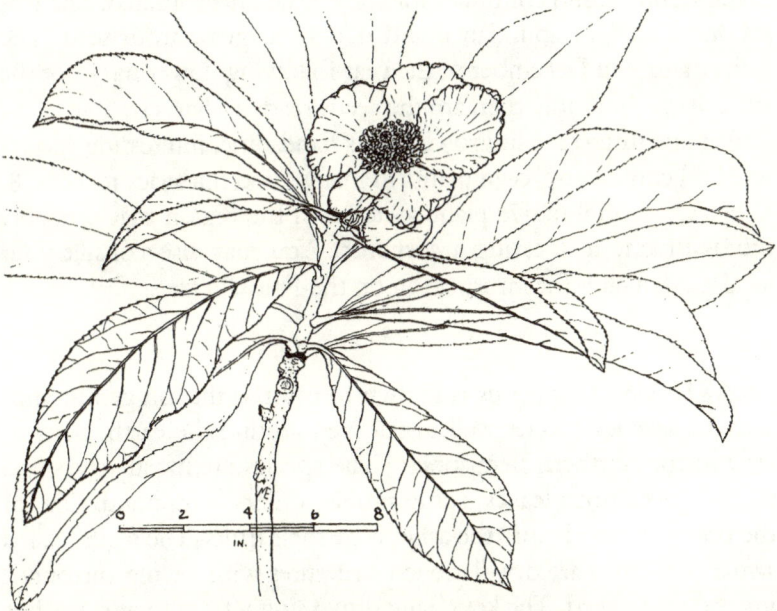

Figure 51. Franklinia alatamaha (Franklin Tree)

has not been found wild since 1803, when it was last discovered by John Lyon, a Philadelphia nurseryman. The species now exists only in cultivation. Though its locality was fairly well defined by Bartram—along the Altamaha River in southeast Georgia, in sandy soils north or east of Ft. Barrington—all efforts to collect the species again have been unsuccessful. Fortunately, Bartram cultivated the plants (or some of his colleagues did) and there are now quite a large number of places where they are grown. Some of the better-known nurseries carry the plant, but care in selecting a site that has light shade or near full sun is important.

The Franklin tree is a large shrub or small tree, up to 30 feet tall. The bark is smooth and either dark brown or blackish. The leaves are wider toward the tip, up to 6 inches long, bright green and glossy on the upper surface, turning scarlet in autumn. They grow clustered on the branches, toward the tips of the season's growth. The flower is about 3 inches across, white, and without any apparent stalk holding it away from the tree branches. Flowering occurs in the summer, and continues for some time once initiated. The tree produces seed in abundance and is easy to grow from seed. The fruits mature in December of the year following flowering—not the same year—and only then are the seeds ready to harvest.

Anyone interested in growing this handsome, interesting species will find container-grown plants in many good nurseries in Zone 8. There are several native plant societies in the region that probably can assist one in locating a specimen. You may also consult your local garden club for information on the Franklin tree.

FRAXINUS. This genus is known to most as the ash genus, with sixty to seventy species well distributed around the earth, particularly in the northern hemisphere. The species are mostly trees and all have deciduous leaves, some simple, others compound. One of the best ways to identify the ashes is by their fruits. The fruits, called samaras or keys, are dry, flattened structures with a wing surrounding the single seed. The keys hang down singly from their stalks but there are many of them in each cluster. They are formed on the second-year twigs, not on the current year's growth.

———*Fraxinus americana,* WHITE ASH (fig. 52). The white ash, one of the tallest trees of the eastern United States and eastern Canada, occurs from Nova Scotia south to northern Florida, mostly east of the Mississippi River. It grows naturally in forests of mixed hardwoods. In Zone 8, it may be found in good numbers in the Naval Live Oaks Area of the Gulf Islands National Seashore Park, where they are mixed with live oaks. They are also present in other areas, of course, but because the trees are very useful for timber, many of the biggest and best have been cut. The plants are tall and slender, with unbranched trunks when forest-grown, but forming a nice, rounded shape with low branches if grown in the open. The compound leaves are deciduous, 8–12 inches long, with the individual

Figure 52. Fraxinus americana (White Ash)

leaflets 3–5 inches long. The samaras (defined above) are slender and light brown or tan at maturity. The white ash is a very satisfactory shade tree and grows fairly rapidly. Its wood is tough and pliable, one of the best for hoes and shovels as well as for furniture making and baseball bats. The white ash is easy to grow, either from seed or from transplants from the nursery. In our soils, it should be given a good start with a mix of peat moss, manure, and soil in equal quantities. Give it plenty of room to grow to get the full benefit of its beauty.

❦

GARDENIA. These plants, as much as the magnolia, are associated with gracious southern living even though they were grown in Europe and originally came from the subtropics or tropics of the Old World. There are about 200 species of gardenias, either shrubs or trees, and perhaps a half-dozen of these are in, or have been brought into, cultivation somewhere. The plants are favorites in greenhouses and other appropriate indoor spaces in climates where it is too cold to grow them outdoors. Gardenias grown in Zone 8 are more or less tender and need protection from the coldest weather. They should be grown in partial shade and will stand deep shade. Under southern pines they get just about the right amount of sun and shade. A good soil or a well-prepared soil mixture, including either peat moss or leaf mold, well-rotted manure, and sandy loams or fine sand in a 1–1–1 proportion is good. Gardenias can tolerate fairly dry weather, but do much better when regularly watered. One of their draw-backs, which can be overcome, is a black, sooty mold that develops on the leaves in late summer. The mold occurs after an attack of white fly, which causes the leaves to exude fluids on which the black mold thrives. A good, general purpose fungal spray used for roses, controls this problem. Control is also achieved by using insecticides to kill the white fly, applied once a week for three weeks, and an oil spray in late fall and early spring to kill the eggs of the insect.

————*Gardenia jasminoides,* COMMON GARDENIA, CAPE JASMINE (fig. 53). You will frequently see the ending "-oides" on many scien-

Figure 53. Gardenia jasminoides (Gardenia)

tific species names. This suffix means "-like," as in "jasmine-like." The botanist who first gave the scientific name to this gardenia thought its flower looked like a jasmine, but he knew it was not a real jasmine. Our plants are evergreen, growing 5–6 feet tall, with summer flowering. The white flowers, most often double, are very fragrant—almost overpowering when a bunch is brought into the house, but very nice outdoors.

———*Gardenia jasminoides*, 'Radicans', DWARF GARDENIA. This dwarf form has flowers quite similar to those of the larger plants, but the leaves tend to be narrower than those of the bigger form. It is subject to the same white fly infestation and is treated similarly. The small plants make nice edging for walks or along patios.

GELSEMIUM. It is strange that there is so much variation in the number of species found in one genus; some genera have hundreds of species, while others have only one, two, or three. *Gelsemium* is a genus of the latter type—only three species, one of these native to Asia and the others to the southeastern United States. Though there are theories about how such a pattern of geographic distribution occurred there is no definite proof, and this is one of the fascinations of the study of plants. All three species are vines, and all three climb by twining around whatever object they find rather than by hanging on with tendrils as do grapes and other vines. The one species well known in Zone 8 is a native with a large area of distribution, from Virginia south and west to Texas and Central America.

―――*Gelsemium sempervirens,* YELLOW JESSAMINE, CAROLINA JASMINE, EVENING TRUMPET FLOWER (fig. 54). These common names are okay, but the authors know this species as YELLOW JASMINE. Perhaps "jessamine" is used to point out that it is not a true jasmine, which is in the olive family. This species is in the logania family, one not too well known to many people. The yellow jasmine is evergreen (*sempervirens*) and blooms for long periods in spring and summer. It blooms most prolifically in open areas but will grow in shade. It is sweet-scented, but one must be careful with this plant for its leaves and flowers are toxic. It is bad for livestock, which will eat it when not much else is left after dry spells. The plant has been implicated in the poisoning of people who (one observer implied) have eaten honey made from the nectar. This was not a documented case, however. One of the authors, D. J. Rogers, had the yellow jasmine in his yard when he was growing up, and he always hoped that it would bloom at the same time as the wisteria planted on the same trellis—sometimes it did, sometimes not. But when it did, it was a joy to behold the bright-yellow jasmine flowers with the lavender clusters of wisteria, and both produced a delightful fragrance.

GINKGO. We have only a few individuals of this wonderful tree (there is one species in the genus, from eastern China), but more

Figure 54. Gelsemium sempervirens (Yellow Jasmine)

should be planted because of the species' very handsome growth pattern. It is a very symmetrical tree, growing 50–60 feet tall (to 120 feet in its native habitat). Although western botanists have known this tree for a long time (it was first named in 1753 by Linnaeus, the most famous of all plant classifiers), its cultivation goes back even farther in the Far East, where its origins are still unknown. There, the only place one finds the trees is in someone's garden or temple-yard. The species has been cultivated all over the world in temperate regions and is now popular as a street tree in the inhospitable environment of large cities.

———*Ginkgo biloba,* MAIDENHAIR TREE (fig. 55). The species gets its common name from the form of the leaves, which much re-

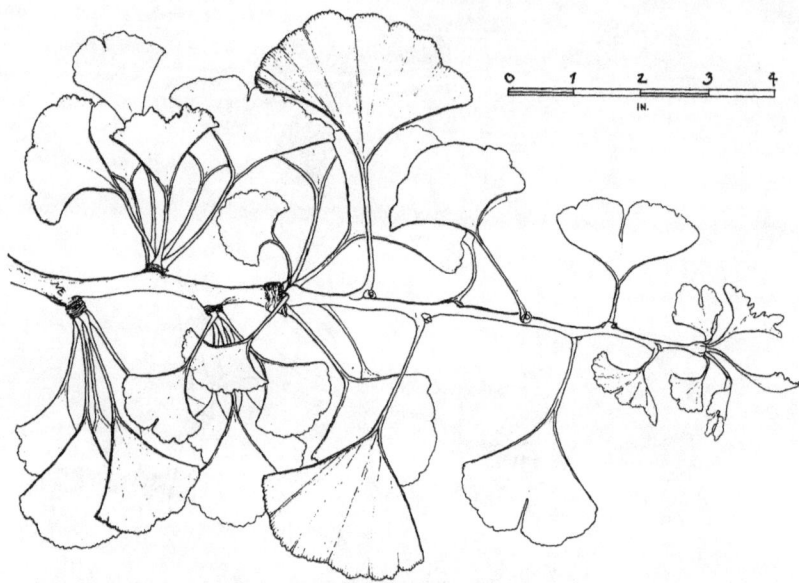

Figure 55. Ginkgo biloba (Ginkgo)

semble those of certain ferns. This is a deciduous tree related, though distantly, to various conifers, such as pine trees. It also has some characteristics of plants lower on the evolutionary tree, thus making it quite a conundrum to those whose job it is to show evolutionary relationships among the plants. The two sexes are borne on separate plants. Only the male trees should be planted because the seeds of the female trees are quite odoriferous, unpleasantly so. Gardeners may ensure the sex of their trees either by using rooted cuttings or by grafting known male scions onto stock of either sex.

❦

GORDONIA. This is another member of the tea family, Theaceae, which includes several genera of horticultural importance; *Camellia* is the outstanding example that comes to mind. Like *Camellia*, *Gordonia* has Asiatic representatives, but unlike Camellia, Gordonia also contains American native species.

———*Gordonia lasianthus,* LOBLOLLY BAY, BAY, BLACK LAUREL (fig. 56). The loblolly bay is native to moist, low areas of Zone 8, where it is mixed with other trees such as silver bay magnolia and

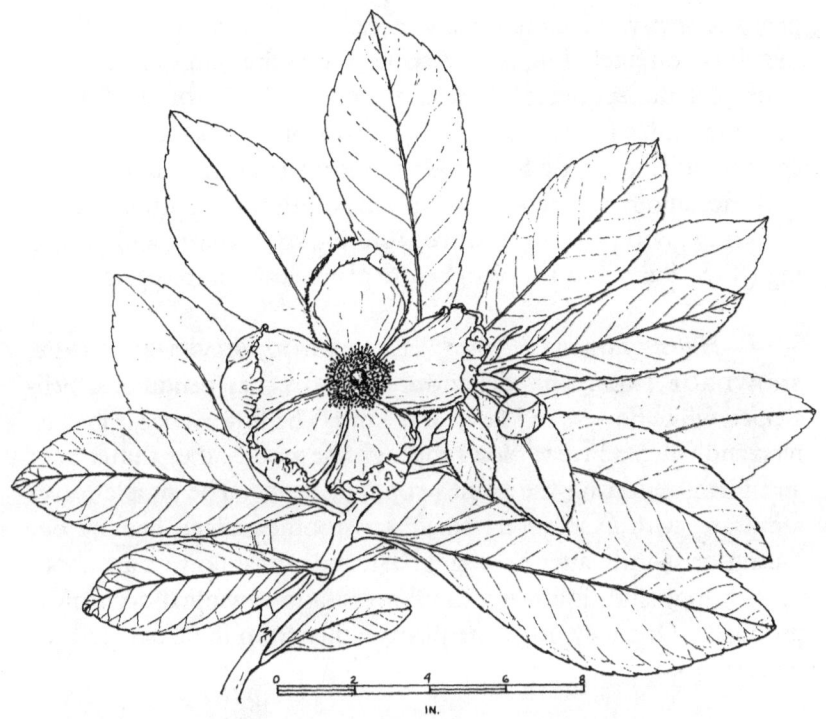

Figure 56. Gordonia lasianthus (Loblolly Bay)

bald cypress et al. This evergreen tree may be rather tall, up to 90 feet, but most specimens are under 50 feet. From a distance, the loblolly bay may be confused with the silver bay magnolia, but at closer range, it can be distinguished by the under side of the leaves: dull, light green for loblolly bay, light gray for the silver bay magnolia. The elliptic, glossy leaves are serrulate on the upper two-thirds of the margins only. The flowers are white and showy, 2–3 inches across, borne on stalks about 2 inches long, appearing in June–July. If you have a low spot where moisture is guaranteed most of the year, these plants will do well, and add great beauty, both in flower and in leaf.

HALESIA. This is a small genus of trees with about five species found as natives in eastern China and the United States. The genus is in the family Styracaceae, or the styrax family. A closely-related genus is *Styrax*, which has many more species, several of which are in cultivation, including one described under the genus name in this book. All the species of *Halesia* are either shrubs or small trees, growing under the canopies of taller trees in woodlands. The two species native to Zone 8 are found in similar habitats, in forests of both deciduous hardwoods and pines. Cultivation is quite simple, requiring good planting practice, watering to establish, and fertilizing (8–8–8 is a good choice) before plants leaf out in spring.

———*Halesia carolina (formerly H. tetraptera)*, SILVER-BELL TREE, SNOWDROP TREE. This delightful small tree has pendulous, bell-shaped white flowers in March and April. The flowers occur in clusters, and when a breeze blows through the woods, they flutter back and forth, reflecting the plant's common name. The simple leaves are ovate, with usually very small, saw-toothed edges, coming out during or slightly after flowering. The individual flowers are about ¾ inch long and, following the flowers, a four-winged dry fruit is produced. One may grow the plants either from the fresh seeds or by cuttings.

———*Halesia diptera*, SILVER-BELL TREE (fig. 57). This species and the one above, *H. carolina*, are somewhat similar in their habit of growth and choice of habitat; *H. diptera* is said to be about 30 feet tall and has a single trunk while the *H. carolina* grows to 40 feet and is frequently multi-trunked. The flowers of *H. diptera* are just a bit larger, as are the leaves. About the most significant difference between the two species is that *H. diptera* has two wings on the fruit rather than four. If you have a good shady area, this or the former species will make a fine addition to your yard.

❦

HEDERA. This is a small group of climbing vines, mostly evergreen, from various parts of Europe, the Canary Islands and Japan. The vines are useful in a number of settings; they can climb on walls and fences or provide a ground cover in shady places. There are many forms of the two species mentioned below. Both species are quite useful in the Deep South.

Figure 57. Halesia diptera (Silver-bell Tree)

————*Hedera canariensis,* ALGERIAN IVY, CANARY IVY, IVY. The best way to tell this species from the one below, *H. helix,* is that *Hedera canariensis* has large, unlobed leaves with red leaf stalks (petioles) and the following species has smaller lobed leaves with green petioles. Both species are commonly called "ivy." This species has to be kept in bounds by regular pruning. It can be used as a ground cover under shade trees, but most of us don't want to make a good hiding place for our slithery friends, the snakes. This merely points up the fact that nearly any plant has some drawback.

————*Hedera helix,* ENGLISH IVY (fig. 58). This is probably the predominant species used in Zone 8, but it is hard to make this judgment unless one is a specialist. There are probably more variants of

82 ❦ *Hibiscus*

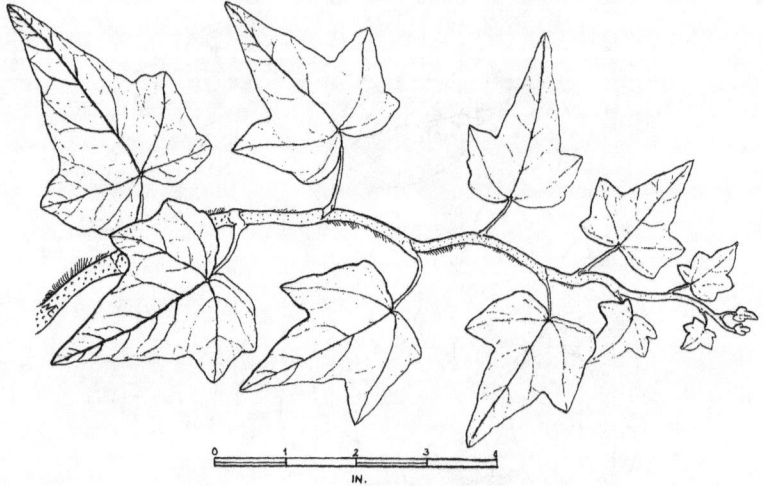

Figure 58. Hedera helix (English Ivy)

the English ivy than of the Canary ivy. An interesting feature of the species is that it has juvenile and adult forms. If one were to plant a seed, one would first get, as expected, the juvenile and then the adult form on the same plant. However, if you start with cuttings of the juvenile form, the plants will remain in that form. This is true of several other kinds of plants as well.

HIBISCUS. *Hibiscus* is a large genus of plants belonging to the family Malvaceae. There are many useful species in other genera in the family, such as cotton and okra, as well as several useful species in the genus *Hibiscus* itself. In *Hibiscus* there is a species grown for fiber (*H. cannabinus*), another for very beautiful wood called "blue mahoe" (*H. tiliaceus*), and several others used for their decorative value. One species in particular—*Hibiscus rosa-sinensis,* the rose of China—is very popular in subtropical and tropical regions, but is rather tender here in Zone 8. Species in this zone include the following.

———*Hibiscus moscheutos,* COMMON ROSE MALLOW, MALLOW ROSE (fig. 59). This species does not really fit with our definition of

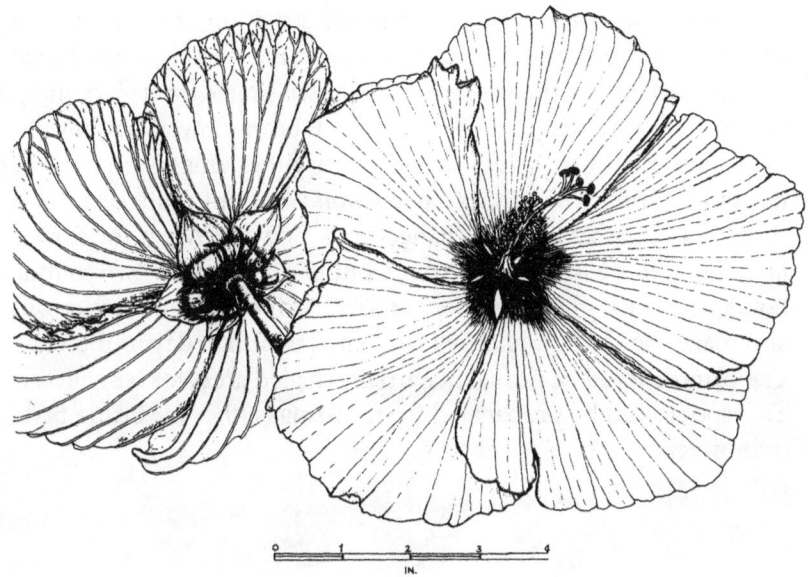

Figure 59. Hibiscus moscheutos (Common Rose Mallow)

woody plants but, rather, is a perennial herb. However, the plant does overwinter with a woody base from which new stems arise each spring. "Author's privilege" is invoked in this case, and it's included. Our garden plants are derived from this species, which is native to Zone 8. The cultivar derivatives have much larger, more handsome flowers. The leaves are broadly ovate with serrate margins, 3–5 inches long and up to 4 inches wide. There are a number of cultivars with white, red, or pink flowers, each flower lasting only one day. Usually there are many buds produced on each of these plants, and a succession of blooms appear over a 2–3 week period. These variations can be grown from seed, but they need extra care in the early stages of growth because the seedlings are very weak stemmed and must be supported for several weeks after they have been planted. However, after the plants pass this early stage, the stems become stronger and stand erect by themselves. No extra care is needed during the winter months but the bases will remain more vigorous in the following season if they have been mulched with pine straw during the winter.

———*Hibiscus mutabilis,* CONFEDERATE ROSE (fig. 60). This large shrub or small tree grows to 15 feet tall and is a very vigorous, hardy species throughout the southeastern United States. In its blooming season, late summer to mid-October, it bears flowers prolifically. Each flower is 3–5 inches across, with either single- or double-flowering forms. The species name *mutabilis* means changeable, and there is a remarkable color change in its flowers. When the flower first opens, it is a lovely, pure white, but in a short time (two to four hours) the color gradually changes to a blush pink. Cut flowers will also make the change, and will last only about one day in the vase. The plants should be grown in partially shaded areas for best effects. This species can be propagated from cuttings either in cutting beds or in water.

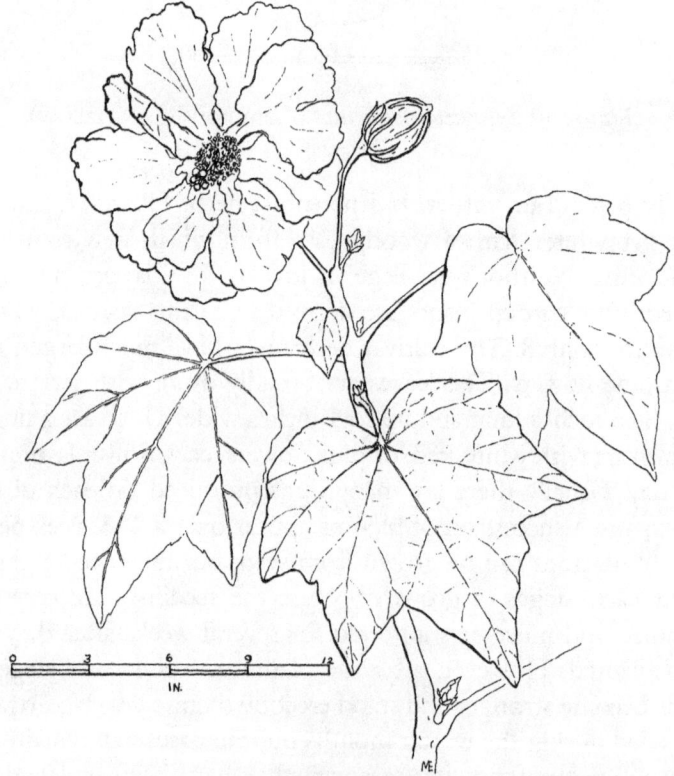

Figure 60. Hibiscus mutabilis (Confederate Rose)

———*Hibiscus rosa-sinensis*, CHINESE HIBISCUS, ROSE OF CHINA, CHINA ROSE (fig. 61). This is certainly one of the most popular of all ornamental shrubs in subtropical and tropical regions. It is marginally hardy in Zone 8 but can be raised if placed in a sunny location on the south side of the house or a brick wall. On the very coldest days, one should cover the plants with a blanket. In case you forget, and your plant gets nipped by frost, just remove the damaged parts, and, usually, the plant will sprout from the base when warm weather arrives. The shrubs seldom grow more than 8 feet tall in our region, but can become much taller in the tropics. They are evergreen with glossy green leaves up to 6 inches long, elliptical to ovate, with serrate margins. The flowers are solitary in the upper leaf axils, and the shrub apparently continues producing flowers all year. Because of their tenderness, we treat our plants as specimens

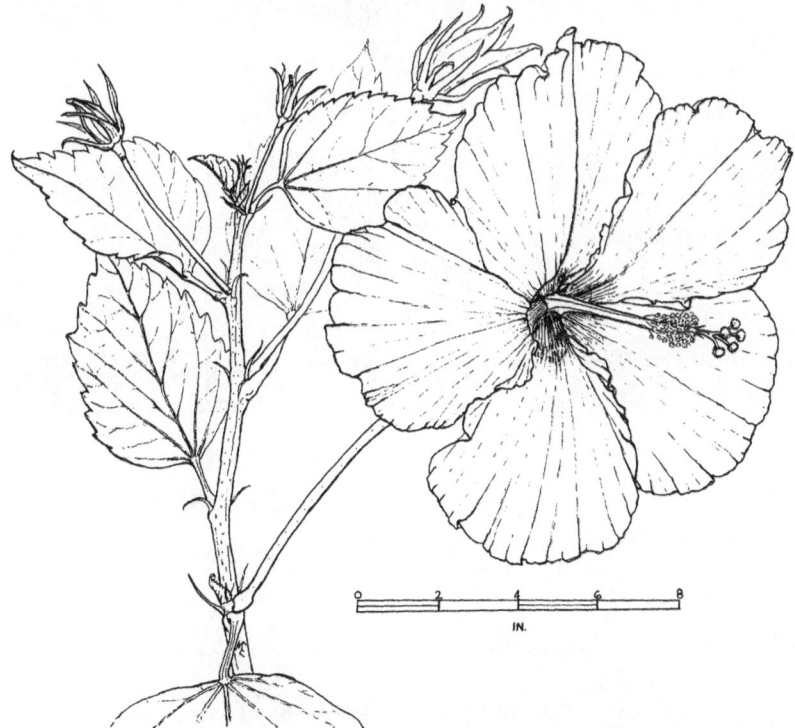

Figure 61. Hibiscus rosa-sinensis (Rose of China)

to be proudly exhibited, but in warmer climes they are used in a host of ways, as basal plantings, as hedges, or as single specimens. There are countless varieties of this species, both single- and double-flowered forms, the flower color of which ranges from white to deep red, yellow, and combinations of these. Other than regular watering, a little fertilizer, and "tender loving care," they need little attention.

————*Hibiscus syriacus*, ALTHEA, ROSE-OF-SHARON (fig. 62). This plant is usually called althea, but in more northern areas the same plant is known as rose-of-Sharon. It is a very vigorous, deciduous shrub growing up to 15 feet tall, in single- or double-flowered

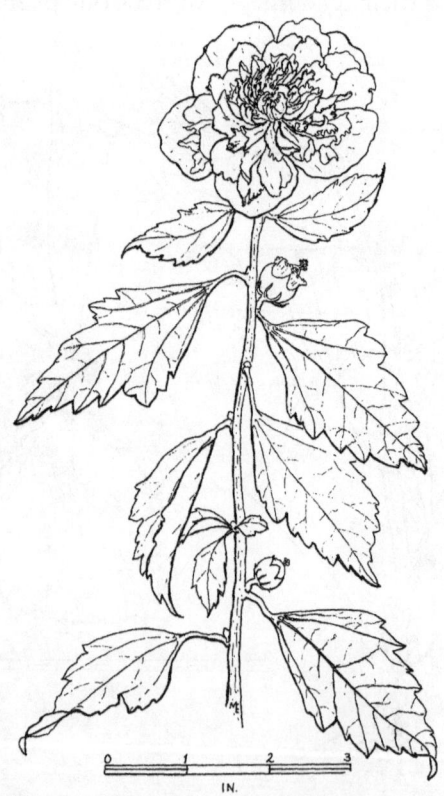

Figure 62. Hibiscus syriacus (Althea)

forms, its flowers white, pink, or magenta, some even a good red. These plants will grow well in a variety of settings but not in deep shade. They are propagated easily by cuttings taken from mature stems and rooted either in cutting media (just sandy soil) or in water. We prefer rooting the cutting in some solid medium because this way its root system is tougher and it's more easily transplanted than a water-rooted cutting. This is one of our most common decorative plants in Zone 8 because of its dependable growth and vigor.

❀

HYDRANGEA. Here, again, are plants whose common and scientific names are the same. This ends confusion because, in this case, the common name refers to all species in the genus. There are several species of *Hydrangea* around the world, but the ones of major interest to us come either from the southeastern United States or from Japan and eastern Asia. These species are mostly shrubs, though one or two show a climbing habit; they are usually deciduous, though if grown in the southern extremities of their ranges they may be almost evergreen. An interesting feature of the plants is that they have two types of flowers: fertile, where seeds are produced, and sterile, with no seeds, and indeed no petals. However, the sepals of the sterile flowers are the large, showy parts for which we raise the plants. These sepals may be all white, all pink, all blue, or in some cases, mixtures of these colors. Variations in soil acidity may cause the sterile flowers to be either blue (acidic) or pink (basic). Not all varieties react readily to change in pH of the soils. Both species of interest to us are natives of North America, east of the Mississippi River.

———*Hydrangea macrophylla,* var. *macrophylla,* HORTENSIA (fig. 63). The common name given is not well known, but some epithet is needed to differentiate this very common garden plant from the next one to be discussed. Hortensia is quite old in cultivation in Zone 8 gardens, being known almost from the earliest days of settlement. It certainly grows well with a minimum of care, making a very satisfactory garden plant that can be used for cut flowers as

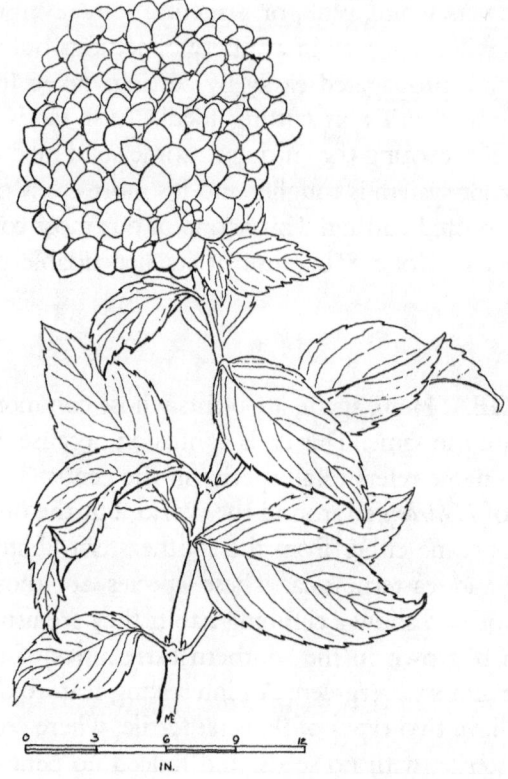

Figure 63. Hydrangea macrophylla var. *macrophylla* cv. Hortensia
(Hortensia Hydrangea)

well. Usually, the bushes are rounded and not often much more than a yard tall, but may be leggier if in good, fertile soils. The rounded flowering heads are composed mostly of sterile flowers, blue in acid soils or pink in basic soils. It is a summer-flowering species. The leaves are large and attractive, and in Zone 8 are mostly deciduous. Very cold days will kill the plant back to its base, but it almost always buds out again in the spring.

————*Hydrangea macrophylla* var. *macrophylla,* LACE CAP (fig. 64). The Lace Cap hydrangea, compared with Hortensia, above, has flowers arranged in flat-topped inflorescences. Instead of nearly all the flowers being sterile, only the peripheral flowers in this variation

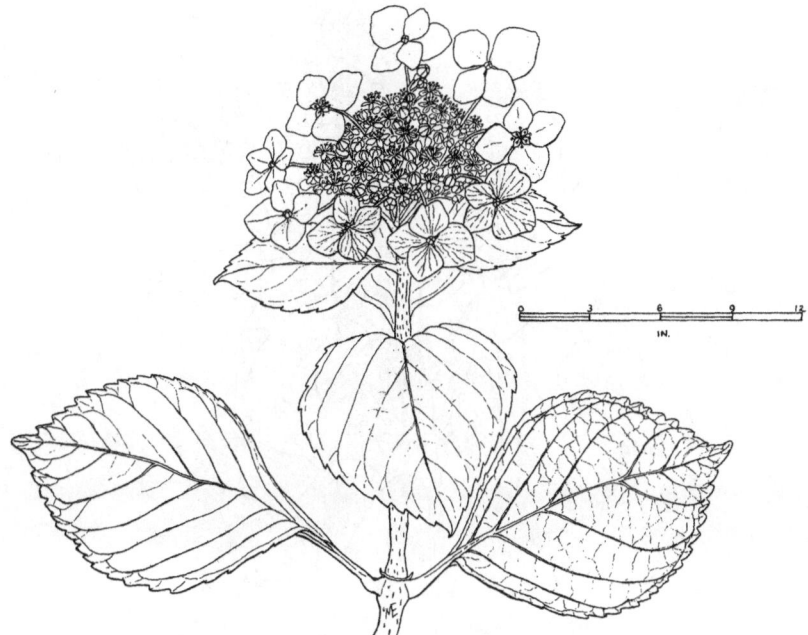

Figure 64. Hydrangea macrophylla var. *macrophylla* cv. Lace Cap
(Lace Cap Hydrangea)

are sterile and the central flowers fertile. This flower form is less frequently seen than is Hortensia. The shape and leaves of the shrub are almost identical to those of the former one.

———*Hydrangea quercifolia,* OAKLEAF HYDRANGEA (fig. 65). This is a wonderful native species found frequently in Zone 8 woods and becoming a major part of the understory vegetation where limestone outcrops occur. One very fine place to see these plants in quantity in the wild is at Florida Caverns State Park, just north of Marianna, Florida. This deciduous shrub has leaves reminiscent of certain types of oak leaves. It is spectacular, or at least handsome, in three seasons: spring, with very nice silvery-green foliage; summer, with large trusses of white flowers; and fall, with pleasing dark-red to russet leaves. The plants have one drawback, and that is their tendency to spread by runners and suckers. The efforts required to keep the plants in bounds is a small price to pay for such a fine

90   *Ilex*

Figure 65. Hydrangea quercifolia (Oakleaf Hydrangea)

ornamental. The species was named by William Bartram, one of our best-known early naturalists.

ILEX This is the genus of holly, about 400 species of trees and shrubs native primarily to the temperate and tropical regions of North and South America as well as Europe and eastern Asia, with both evergreen and deciduous species. Many of the species have pollen-producing male flowers on one plant and fruit-producing female flowers on another (*dioecious* in technical terminology). Our own native species are evergreen with handsome leaves that are small or large, smooth-edged, saw-toothed, or thorny, and most frequently quite stiff. Many dwarf forms have been brought into cultivation or developed by breeding. Thus the genus contributes in many ways to yards and gardens: as individual specimen trees, as shrubs for borders and hedges, or as accent plantings with deciduous shrubs.

──────*Ilex* × *attenuata* cv. East Palatka (fig. 66). This recently introduced cultivar is a hybrid between *Ilex opaca* and *I. cassine*. It is a handsome, evergreen tree 30–50 feet tall. The leaves usually have either just one terminal spine or only a few spines terminally and laterally.

Figure 66. Ilex × *attenuata* cv. East Palatka (Holly)

———*Ilex cornuta,* CHINESE HOLLY (fig. 67). This species, a small tree or large evergreen shrub, is very frequently used in horticulture in one of its many forms. Two of these are listed below.

———*Ilex cornuta* cv. Burfordii, BURFORD HOLLY (fig. 68). This is a handsome, shrubby plant (or sometimes a tree to 12 feet tall) with tremendous numbers of large, orange-red berries, and smooth-edged (almost spineless) glossy leaves.

———*Ilex cornuta* cv. Rotunda. This variation of *I. cornuta* is a very handsome dwarf with closely spaced branches and leaves. The leaves are very spiny. The plants are apparently sterile and make an excellent hedge or foundation planting.

Figure 67. Ilex cornuta (Chinese Holly)

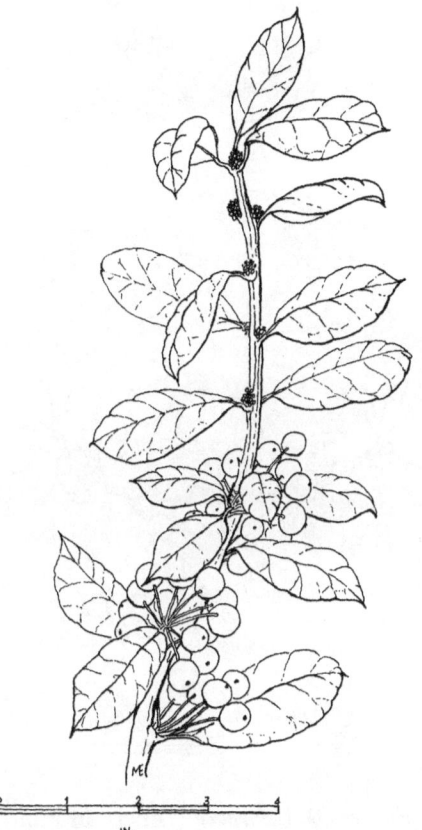

Figure 68. Ilex cornuta cv. Burfordii (Burford Holly)

————*Ilex opaca*, AMERICAN HOLLY (fig. 69). This native grows commonly in Zone 8 forests and has been frequently introduced into yards, where it will grow up to 50 feet tall with a trunk of 8–10 inches in diameter. The bark is a handsome mottled gray and smooth. Branches usually grow straight out from the trunk, but can be ascending or drooping. The thick, leathery leaves are 1–3 inches long and up to 1 inch across, thorny, and not shiny. The flowers are carried in the angles between stems and leaves for some distance back from the tip of the branch and are inconspicuous and white. The bright red berries are familiar to all and add much brightness to Christmas decorations. The berries are also often eaten by birds,

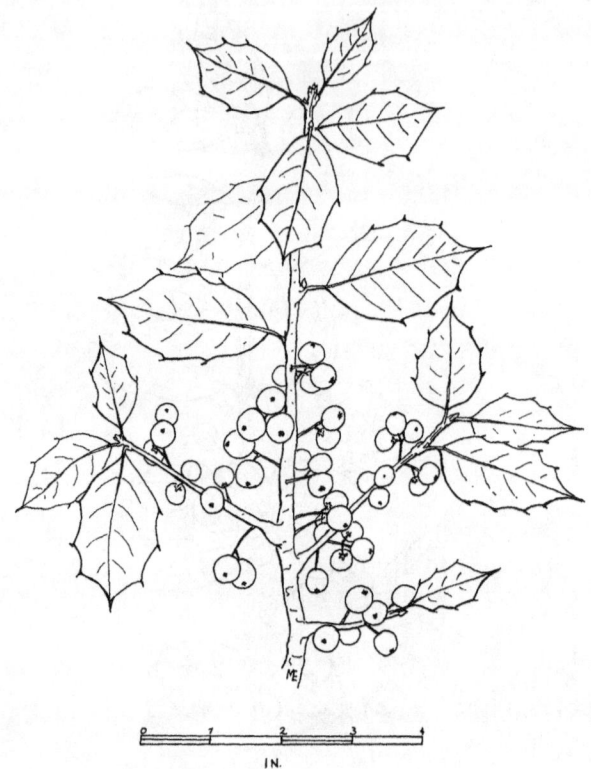

Figure 69. Ilex opaca (American Holly)

frequently serving as food for some of the more interesting migrating species. Cultivation of this species is easy but, as with any plant, care in placing the plants, watering in dry spells, proper pruning to shape, and addition of fertilizer at the beginning of the growing season and again at flowering time will greatly improve the vigor and showiness of the plants.

————*Ilex vomitoria,* YAUPON (fig. 70). This is a large shrub with small, minutely scalloped, evergreen leaves and light-gray bark. Its flowers are indistinct but its fleshy fruits are a handsome sight in the late fall and winter, bright red (infrequently a yellow-fruited form may be found) with many clustered along the stems. Yaupon sprigs are used in Christmas decorations, but the yellowish spots the fleshy

Figure 70. Ilex vomitoria (Yaupon)

berries leave when crushed are not much appreciated on modern wall-to-wall carpeting. A dwarf variety is now much used in hedges where it fills much the same role as boxwood hedges. Yaupon has the merit of growing more vigorously in sandy soils than does the boxwood. The species name *vomitoria* derives from the practice in early times of making tea from the leaves to induce vomiting.

ILLICIUM. This genus of evergreen shrubs has species in both Asia and America. The species described below prefers to grow in moist, shady regions with muck soils, but has been grown successfully in yards with sandy soils and no more than usual available water.

―――*Illicium floridanum*, FLORIDA ANISE (fig. 71). This native species is sometimes also known, inelegantly, as "stink bush." The leaves of the Florida anise are dark glossy-green above, lighter beneath, evergreen, smooth-edged, elliptic, have a pungent odor when crushed, and are about the same size as those of the silver bay, about 5–6 inches long. The shrub, if grown without crowding, will develop a nice, rounded shape about 8–10 feet tall. The flowers, two or more carried together, are nodding and are dark red to purplish, about 1 ½ inches across, with many linear petals. According to one book, the flowers "have the odor of decaying fish," a characteristic that certainly would account for the common name noted above.

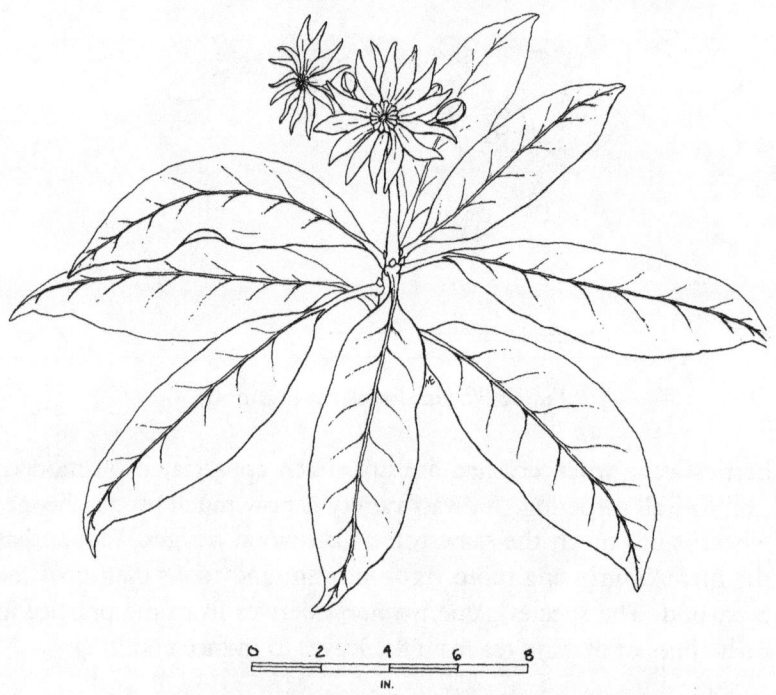

Figure 71. Illicium floridanum (Florida Anise)

However, not everyone notices this odor in the flowers, thus there may be some variation in the plants from one place to another. The Florida anise is another Zone 8 native and is found only from west Florida (and probably the southern counties of Alabama) west to Louisiana. Though most of the books cited in the Selected Reading List mention these plants, there is little information about cultural practices such as fertilization, insect pests, diseases, or pruning.

❀

JASMINUM. There are many species of the jasmines from the tropics or subtropics of China, southern Asia, Africa, the Canary Islands, and Australia. Only three or four of these are commonly found in Florida, and only one occurs in Zone 8 gardens. There are several species in other genera and families that are called jasmine, such as Cape jasmine, yellow jasmine, or crape jasmine, but these are not closely related to the real jasmines.

————*Jasminum mesnyi*, JAPANESE JASMINE, CHINESE JASMINE, PRIMROSE JASMINE (fig. 72). Chinese jasmine is a shrub, up to 10 feet tall, with arching or drooping evergreen stems. The younger stems are square. The leaves are compound with three leaflets, each leaflet about an inch long. The yellow flowers are borne singly in the axils between the stems and leaves along the upper branches. The major attraction of this species is that the plants are evergreen. The flowers make a nice accent but are not too showy. The plants

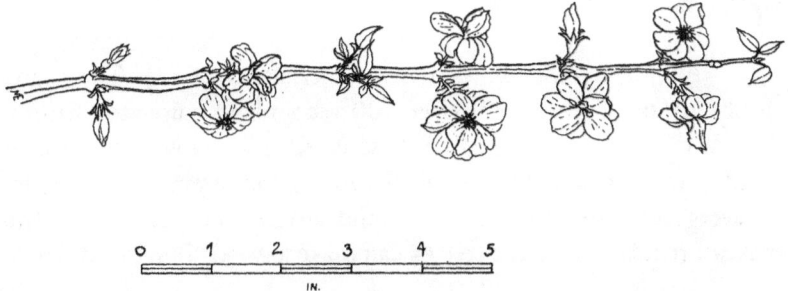

Figure 72. Jasminum mesnyi (Japanese Jasmine)

grow well in sun or shade and thrive in sandy soils. They are useful in borders or low hedges.

❈

JUNIPERUS. This is a genus of evergreen trees and shrubs, both erect and prostrate, or decumbent. The plants are native to North America, indeed to the whole Northern Hemisphere, from the arctic to the tropics. The red cedars have small, scale-like leaves on the mature stems and small, needlelike leaves on the immature stems. Other species may have only small, appressed, scalelike leaves, and still others only needle-like leaves. Many of the species are used horticulturally, and some are used for their wood, as in red cedar closets, wardrobes, clothes chests, pencils, and other items where the reddish wood is decorative. The famed European juniper provides "berries" that contain an oil used in flavoring English gin. As in many other cultivated evergreen trees and shrubs, there are dwarf forms that are very attractive and frequently used. If we had to pick the most commonly used genus in gardens and yards, we are sure it would have to be the junipers. The genus includes perhaps 70 species in all, of which those most commonly found as ornamentals are *Juniperus chinensis, J. communis, J. conferta,* and *J. virginiana.*

Since all of the above have been in cultivation for a long time, it is not surprising that there are many different cultivated varieties in each of the species. One should consult one's nurseryman for those most used in this area. The foliage may be deep green, greyish green, bluish, yellowish, or variegated green with white according to species and form. Junipers are easy to grow and require little attention.

————*Juniperus chinensis,* CHINESE JUNIPER (fig. 73). These are shrubs or small trees with over 100 recognized cultivated forms. One of the most interesting of these is 'Kaizuka,' a very handsome slender tree whose branches spiral upward, each branch resembling an evergreen Christmas tinsel, wound around the tree. This form makes a handsome accent tree or can make a good "boulevard" tree.

————*Juniperus communis,* COMMON JUNIPER. This is one of several species with a prostrate or creeping type of growth. These plants

Figure 73. Juniperus chinensis cv. Kaizuka (Chinese Juniper). Plant 15 ft. tall.

are very useful as ground covers, on banks or slopes, and, in some cases, as foundation plantings.

―――――*Juniperus conferta,* SHORE JUNIPER (fig. 74). This is another prostrate species imported from Japan that is now frequently found in nurseries. Its leaves are needle-like. While the main stem is horizontal, the side branches stand more or less erect and may grow to more than 1 foot tall. The shore juniper is perhaps one of the most vigorous species and seems to thrive in the harshest environmental conditions.

―――――*Juniperus virginiana,* EASTERN RED CEDAR (fig. 75). Some authorities in this region (Clewell 1985, Godfrey 1988) prefer to call our red cedar *Juniperus silicicola* but we prefer to keep the species under the larger umbrella of *J. virginiana.* This is the tree native to Zone 8, occurring northward at least into Virginia and indeed far-

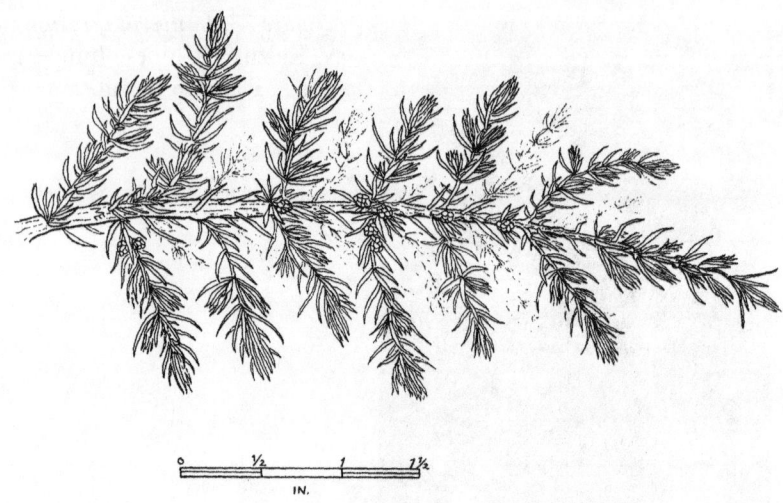

Figure 74. Juniperus conferta (Shore Juniper)

ther north. Normally the red cedar becomes a small tree, and some of the older ones will be 8–10 inches in trunk diameter. Many interesting forms have been developed and used to meet various types of landscaping requirements. One may prune the red cedar into fanciful shapes (topiary) or use the plants as a hedge, wind-break, or accent plant. Many are transplanted from the wild, but to insure its survival, you should buy your plants from a nursery. Young plants may also be bought in large numbers from the county forester. This species grows rather rapidly given good care and fertilizer.

❦

KALMIA. A genus of only 6 species, and these are evergreen shrubs native to North America and Cuba. Our species is found in deep woods with fairly rich, acid, sandy soils. The flower is shaped like a wide-open cup, has a fringy edge, and is white turning pink or pinkish in early spring, usually about March or April. The flowers are clustered at the ends of branches in umbrella-shaped (umbelliform or corymbiform) groups. The foliage is poisonous if eaten.

———*Kalmia latifolia*, MOUNTAIN LAUREL, CALICO BUSH (fig. 76). The species name means broad leaf. This is a very widespread spe-

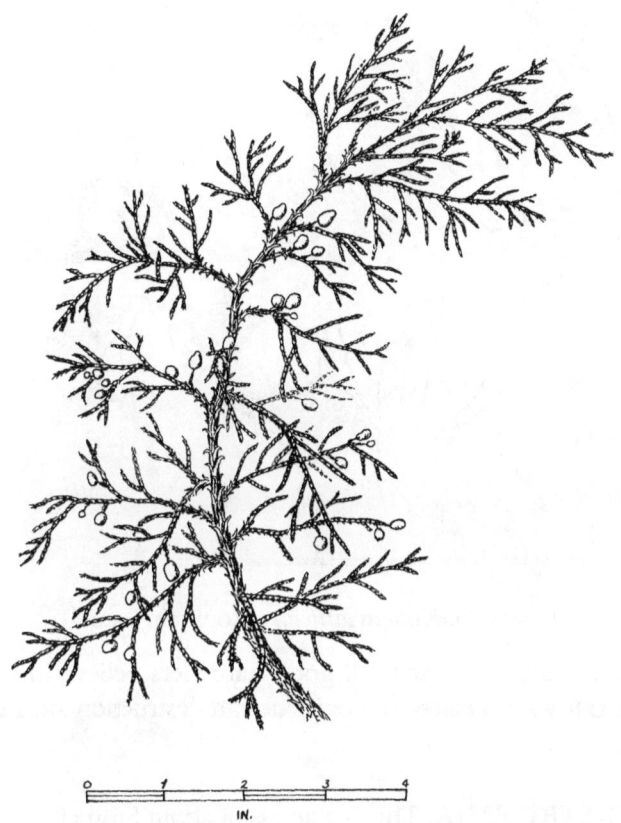

Figure 75. Juniperus virginiana (Eastern Red Cedar)

cies, occurring up and down the Appalachian Mountains. Zone 8 is just about at the southern end of the distribution in North America. This shrubby species has sharply angular stems that are 12–15 feet tall under ideal conditions. The flowers are generally found in terminal, umbrella-shaped clusters, many whitish flowers appearing in March and April and turning pink with age. The plants prefer good, rich soils, high in organic matter and a bit on the acid side. *Please* do not dig up the wild plants! More often than not, such mature plants will die when brought in because of broken root systems. Get plants from a nursery: they will be better plants because the roots have been trained and clustered so that, when transferred to your garden or yard, the plants will be much more likely to succeed. Con-

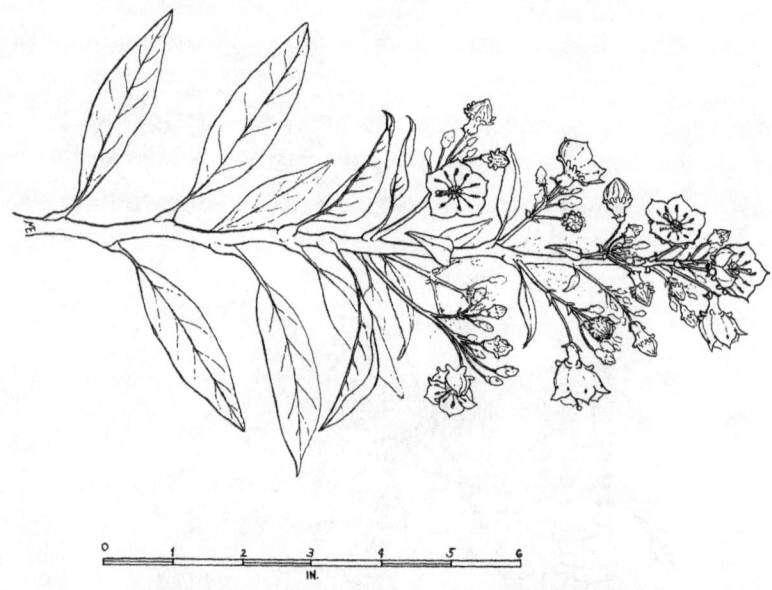

Figure 76. Kalmia latifolia (Mountain Laurel)

servation is a practice that all good gardeners believe in. Do not destroy our wild species—there is enough destruction already.

LAGERSTROEMIA. This is a genus of about 55 species, all from the warmer regions of Asia and the Pacific Islands. Only one of them (discussed below) is adapted well to the southeastern United States. The flowers of all species in this genus are showy, and some of the other species not cultivated in Zone 8 are grown in greenhouses in the North. The species discussed here is useful in many settings, either as a specimen plant, in hedges, grown as a shrub or tree. Cultivation is simple and there are few problems with the plants, though they do tend to have an aphid that produces a sticky substance that causes the leaves to turn blackish. This is easily remedied, however, with common insecticides. Powdery mildew, another problem, can be prevented with a lime-sulfur spray in early spring. Pruning back the plants will cause them to bear more flowers and prevents them from becoming "gawky." (We can't think of an adjective that better describes the older, un-pruned plants.)

———*Lagerstroemia indica*, CRAPE MYRTLE (fig. 77). The crape myrtle is one of the oldest and best-loved species in the South. Just about everyone has a few plants, grown either as small trees or as shrubs. There are some dwarf forms that never grow more than about 3 feet tall, and these make excellent hedges. The large flower bunches—panicles—make a beautiful show, flowers of the various forms ranging from dark, watermelon red to pink to white. Some flower forms are said to be purple, but it's more of a magenta. The petals are interesting, each held on a slender base expanding to a rounded, frilly-edged part above. The leaves of these deciduous plants are rounded and are a nice glossy green. As indicated above, they are easy to grow though they do have a tendency to spread

Figure 77. *Lagerstroemia indica* (Crape Myrtle)

from underground runners. The plants flower in mid- to late summer, when few other species are flowering.

❊

LANTANA. This genus of shrubby species belongs to the verbena family, which is characterized by square stems and opposite or whorled leaves. Members of this genus are frequently prickly and almost all of them have a rank odor. In the tropics, these plants are considered serious weeds, particularly in cattle-growing areas, because cattle frequently die from poison in the plants.

————*Lantana camara,* YELLOW SAGE, LANTANA (fig. 78). This shrub, which grows to about 4 feet tall, is a naturalized plant all

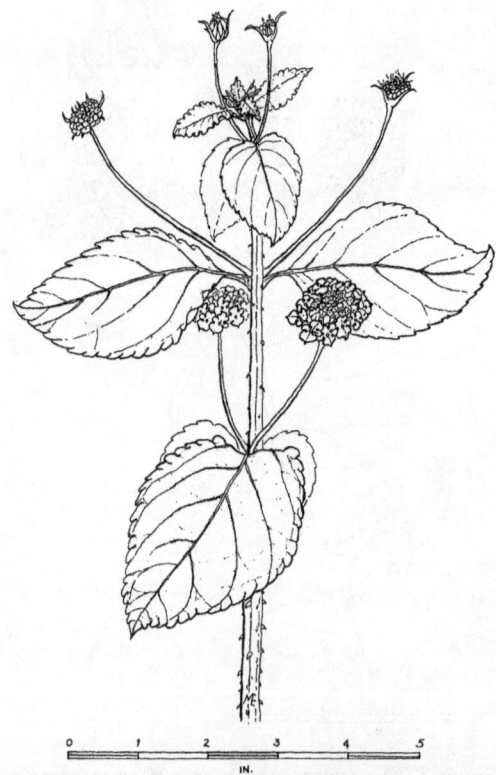

Figure 78. Lantana camara (Lantana)

over the South. It volunteers everywhere and often isn't very well thought of, but it can make a nice showy plant where just about nothing but prickly-pears will grow. The flower head, which is either flat-topped or rounded, is composed of numerous small flowers. There is usually a mixture of flower colors in the same head, yellow, red, and white. Lantana flowers right through the summer. More shapely plants can be produced by removal of dead flowers, which encourages more branching.

❀

LEUCOPHYLLUM. This is a small genus, ten to twelve species, all native to the drier areas of west Texas, New Mexico and bordering Mexico. Two of the species are mentioned in *Hortus III* (see "Selected Reading" list) as being of horticultural quality. All of the species exhibit characteristics of plants from more arid zones: small leaves covered with fine hairs that help prevent too much evaporation through the pores in the leaves. At first it may seem strange that our region, with much higher rainfall, would be a suitable environment for plants from such a dry region. But our sandy soils, which lose water very rapidly, are actually much like those of arid regions; witness, for example, the century plant from the dry Southwest, which thrives in Zone 8.

————*Leucophyllum frutescens,* TEXAS SAGE, CENIZA, BAROMETER BUSH (fig. 79). Texas sage is an evergreen shrub with many branches set close together, giving it a compact appearance. We have not seen enough of these plants to get much of an idea about the maximum height the plants may attain. The ones we have seen were not more than 4 feet tall, but these were recent plantings and mature plants may double that height. We believe that this species has been relatively recently introduced into our gardens. The gray-green leaves covered with silvery hairs are the most striking feature of this species, though the hairs are difficult to see without magnification. The flowers are a rosy-pink, about an inch across, borne in great numbers and producing an attractive effect from late summer to early fall. These plants provide a handsome accent, either singly or planted in twos or threes in an island surrounded by a low border,

106 ❦ *Leucothoe*

Figure 79. Leucophyllum frutescens (Texas Sage)

such as monkey grass. They should be planted in full sun or only partial shade.

❦

LEUCOTHOE. Plants of this genus are generally called "fetterbush" because the bushes generally have flexible stems that horsemen could use to tie the front legs of their horses to keep them from wandering too far at night, thus fettering them. We have not seen any individuals of the species native to Zone 8, *Leucothoe axillaris,* grown as horticultural materials in gardens, though there may be some. The species discussed below is native to areas a bit north of Zone 8. Selections of this species have been made and are sold by nurseries. We have one of these in our yard and think it is worth the effort.

———*Leucothoe populifolia*, FETTERBUSH. The typical plants of this species grow 10–12 feet tall, with bright, green, evergreen foliage. One usually plants a single specimen for accent at corners of the house, or as the center around which other, lower-growing species may be planted. The stem and branches are quite compact, making for rather dense growth. The individual leaves are 3–4 inches long, more or less elliptic, with finely serrate (sawtoothed) margins and a slender, pointed tip. The plants are subject to both insects (leaf miners) and fungus (round blotchy spots), which may be controlled with any good general insecticide and fungicide. The flowers occur in mid-spring (April) in racemes. Each flower is urn-shaped, very typical of the blueberry family, Ericaceae (also sometimes called the heather family). The plants do well in acid soils and improve with a mulch of pine needles.

LIGUSTRUM. There are about fifty species in this genus, most of which are from eastern Asia. Zone 8 species are evergreen, but there are some others that are deciduous. These plants are mostly shrubs, but some, if left to grow, will become small trees. All are easy to propagate from cuttings, from seeds, or by grafting. The small black fruits of the species listed below resemble a miniature olive—the plants are members of the olive family—and are food for birds. However, if one wants to have well-formed plants, pruning to remove both flowers and fruits will be necessary. The small, white flowers grow in clusters and are usually sweet scented, but the plants are raised mostly for their foliage, in hedges or background plantings. Cultivation is very simple, and the plants grow without much care.

———*Ligustrum japonicum*, JAPANESE PRIVET, WAX-LEAF PRIVET, LIGUSTRUM (fig. 80). This shrub, if left untrimmed, will grow to be 10 feet tall. It has been grown for a long time in the Deep South. Its oval, glossy, evergreen, thickish leaves make it a very good foun-

108 ❦ *Ligustrum*

Figure 80. Ligustrum japonicum (Wax Leaf Privet)

dation plant. It has been confused frequently with the following species.

————*Ligustrum lucidum,* GLOSSY PRIVET, CHINESE PRIVET. This is a larger species than the above, up to 30 feet tall. Modern landscape architects frequently use it around buildings in malls or other large structures where they take advantage of the glossy foliage (it has larger leaves than the preceding species), and the striking, light gray branches exposed by pruning below the upper leafy portion.

————*Ligustrum sinense,* COMMON PRIVET, HEDGE PLANT (fig. 81). This is probably the most commonly used hedge plant. It is vigor-

Figure 81. Ligustrum sinense (Common Privet)

ous and, with pruning, thick and well-leafed from top to bottom. This privet has become naturalized in Zone 8 and is such a vigorous grower that it will crowd out other native plant materials. It also has the disadvantage of harboring clouds of white fly, a pest that causes problems on many other plants, such as roses and gardenias. Frequent spraying of the hedge plants is required, but if there is a brushy area nearby with the volunteer plants growing, it is nearly impossible to keep the pest down.

LIQUIDAMBAR. There is but one species of this genus of trees in the United States. Two others are found, one in Asia Minor, the other in Taiwan (formerly Formosa). All are trees. The liquidambars produce a valuable timber and an aromatic balsam, called storax or styrax, important in medicine and perfumery.

———*Liquidambar styraciflua,* SWEET GUM (fig. 82). This handsome, native, deciduous tree, which may grow to be 120 feet tall, is valuable both as a shade tree and for its lumber. As the scientific name implies, it exudes a gum that can be used as mentioned above. The plants grow well and easily transplant from the wild or come up as volunteer plants. The leaves, with saw-tooth margins, are five- to seven-lobed and are reminiscent of maple leaves. These are glossy-green in summer and frequently turn a bright yellow or a

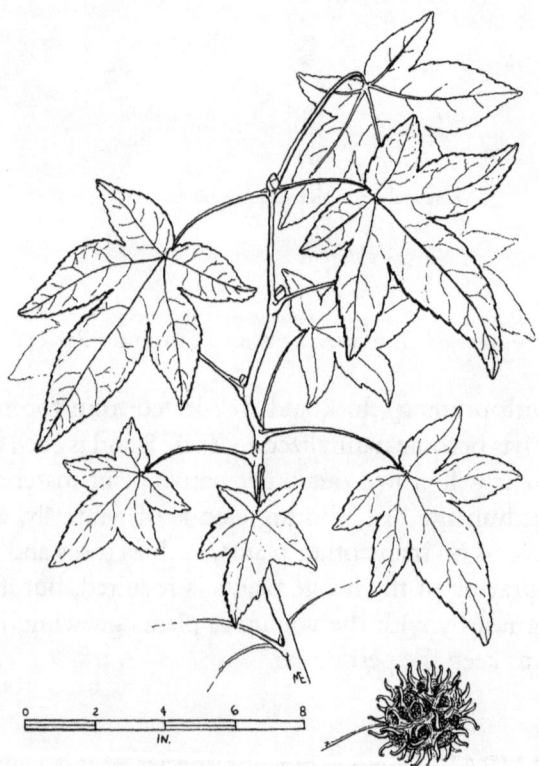

Figure 82. Liquidambar styraciflua (Sweet Gum)

deep red in the fall. Some selections of the trees that have more guaranteed fall color than a chance wild seedling may be found in nurseries. Perhaps the greatest drawback to use of this plant in yards is the fact that it produces rounded, burr-like fruits in profusion, which drop at maturity and make for a messy appearance.

❊

LIRIODENDRON. The Deep South's native tulip tree or, as it is better known among lumbermen, yellow poplar, is one of only two species in this genus. The Zone 8 native occurs through much of the eastern United States. The other species comes from central China. This pattern of geographic distribution, where a genus is shared among China, Japan, and eastern North America, is one that is common to many different plant groups. Species found in the Far East are often duplicated or have closely related species in eastern North America. This puzzling distribution has been recognized by botanists for over a hundred years, and might be explained by the fairly recent geological understanding that the continents, once together, have drifted apart, and their former unity accounts for the botanical phenomenon. However, there are problems in the timing of the events since the continents began separation long before many plants, such as the one discussed here, were thought to have evolved.

————*Liriodendron tulipifera,* TULIP TREE, YELLOW POPLAR (fig. 83). This deciduous tree is in the same family as the magnolia. Their relationship can best be appreciated through close comparison of the flowers of both plants. The shape of the tulip tree flower is reminiscent of a tulip. The petal color is an unusual combination of green and yellow. Our main source of information for writing this account comes from the reference book *Hortus III* (see "Selected Reading" list for further details). This book refers to the tulip tree as one of our noblest trees. In nature, it grows to well over 100 feet tall, may have a trunk diameter of 3 feet or more, and is beautifully shaped. Its leaves are deciduous with a distinctive shield shape, and

112 ❦ *Lonicera*

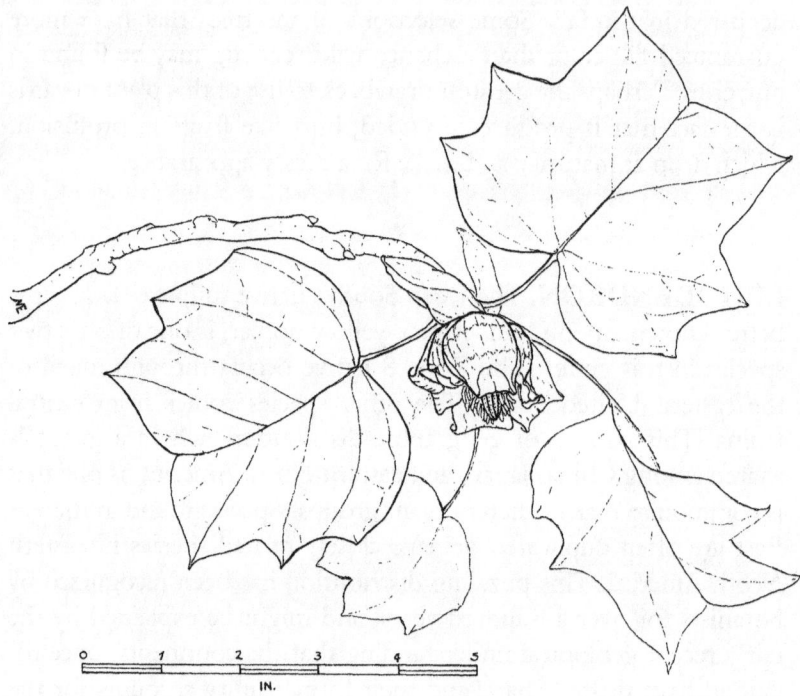

Figure 83. Liriodendron tulipifera (Yellow Poplar)

measure up to 5 inches across. It is said that transplanting this species is difficult, but we have had no trouble taking rather small plants from the wild and moving them to another location. Just don't try it with plants over 1 foot tall!

LONICERA. There are both vines and shrubs in this genus, some deciduous, others evergreen. Some are handsome shrubs, but the two species noted here are vines. The first is a vine that is almost as bad as the kudzu insofar as covering acres of land. The second species stays wherever it has been introduced. Both are called "honeysuckle" because they have nectar. As kids, we used to pull off the petals and suck out the small amount of sweet-tasting juice at the base. These species also have a very pleasing odor.

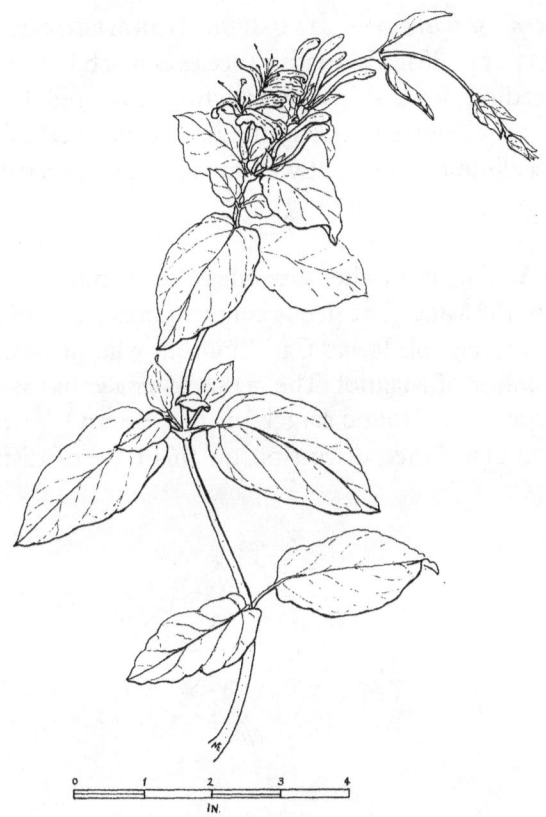

Figure 84. Lonicera japonica (Japanese Honeysuckle)

————*Lonicera japonica,* JAPANESE HONEYSUCKLE, HONEYSUCKLE (fig. 84). This vine, an evergreen, was introduced many years ago, but the introduction backfired. What appeared to be a very nice vine, easy to grow and sweet-scented, turned out to be so vigorous it could spread uncontrollably and it is now a serious weed. You can clear it off your property, but it will soon be back because the birds like the fruits and scatter the seeds everywhere. Even if this vine didn't reproduce by seed, it would still succeed because it produces a mat of underground runners. Perhaps we shouldn't include this species as a "preferred" plant, but it is attractive and the birds do get some food from the berries.

114 ❦ *Magnolia*

———*Lonicera sempervirens*, TRUMPET HONEYSUCKLE, CORAL HONEYSUCKLE (fig. 85). This native species is much better behaved than the preceding one and is very attractive. It has red flowers and a good cultivar (cultivated variety) has been developed called "Superba". It is a climbing vine, but will stay in its place pretty well.

❦

MAGNOLIA. This is another case where the common and scientific names are the same. The genus commemorates an early French botanist, Pierre Magnol. It was Carl Linnaeus who first established *Magnolia* in honor of Magnol. The genus *Magnolia* has as many as 85 species worldwide, found largely in the eastern United States and in eastern Asia. (See discussion on this type of distribution

Figure 85. Lonicera sempervirens (Trumpet Honeysuckle)

under *Liriodendron*.) Several species occur in Mexico, one of which, *M. schiediana*, is very similar to the Deep South's *M. grandiflora*. The Deep South is well endowed with *Magnolia* species, both native and introduced. There are very few other genera that provide so many delightful species in one area. As is obvious from the amount of space we have devoted to magnolias, we have a special fondness for them. And there is no need for excuses, considering the great beauty these plants provide. The discussion below arbitrarily divides the introduced species from the native species.

The native species—the first two with deciduous leaves, the next two evergreen.

————*Magnolia ashei*, ASHE'S MAGNOLIA, CUCUMBER TREE (fig. 86). Ashe's magnolia is a shrubby species, seldom more than 12–15 feet tall. As a native it grows in forest regions as an understory plant, but it will grow quite well in open areas with full sun or light shade. Its deciduous leaves are very large, up to 2 feet long and 10 inches across. They are very handsome, light green, with a silvery appearance underneath that shows up quite well when the wind blows. The large flowers (6–10 inches across) are white with purple dots inside the "cup" of the tepals (technically not petals). They are very attractive, have a very sweet, magnolia-type fragrance, and are quite similar to *Magnolia grandiflora* flowers. This species, which is restricted in its Florida distribution mostly from Santa Rosa County east to Gadsden County in the Panhandle (though it also occurs in Texas), is found infrequently in yards and gardens but certainly deserves more attention from garden enthusiasts. The plants can be grown in more northern regions than Zone 8, as has been shown by Dr. Frederick Meyer of the National Arboretum in Washington, D.C., who has successfully grown them in his own garden for a number of years. Again, people are cautioned against trying to take these plants from the wild because their roots seldom grow close to the base of the stem. The plants can be grown from seeds planted very soon after they ripen, and they will germinate the following spring. It is best to purchase small, living plants from a qualified nursery.

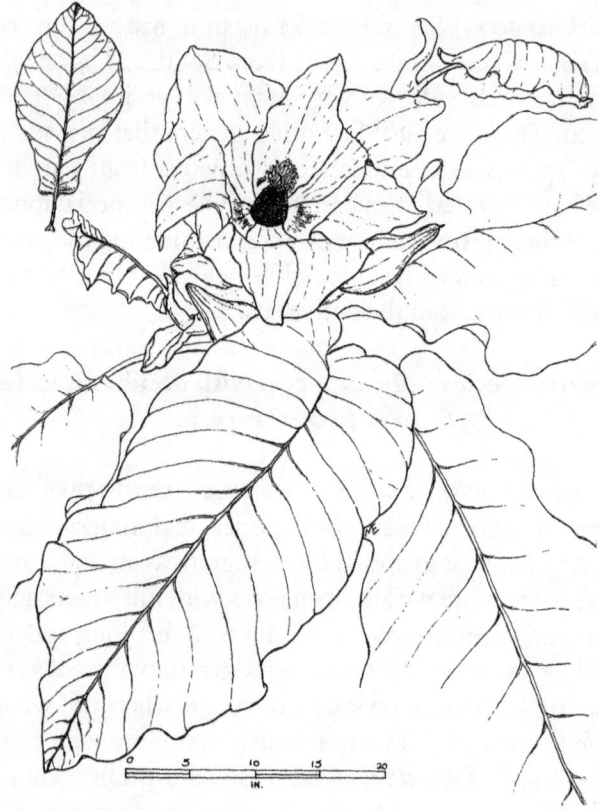

Figure 86. Magnolia ashei (Ashe's Magnolia)

————*Magnolia pyramidata* (fig. 87). There is no common name for this little-known species, but it should attract more attention horticulturally. It has very attractive, distinctive foliage and large, white flowers with a delightful fragrance. The plants will grow to heights of 30 feet, perhaps more, in mostly light shade. Once established, they require little attention and can withstand either heavy rainfall or long, dry periods. The leaves grow in tiers, in whorls, with distinct space on the stems between the whorls, giving the plants a very interesting look. The flowers, while smaller than those of the southern magnolia, are about 5 inches across, their petals (tepals) narrower than those of Ashe's magnolia, flowering from late April into May. The fruits take a long time to mature. The seeds pop out of

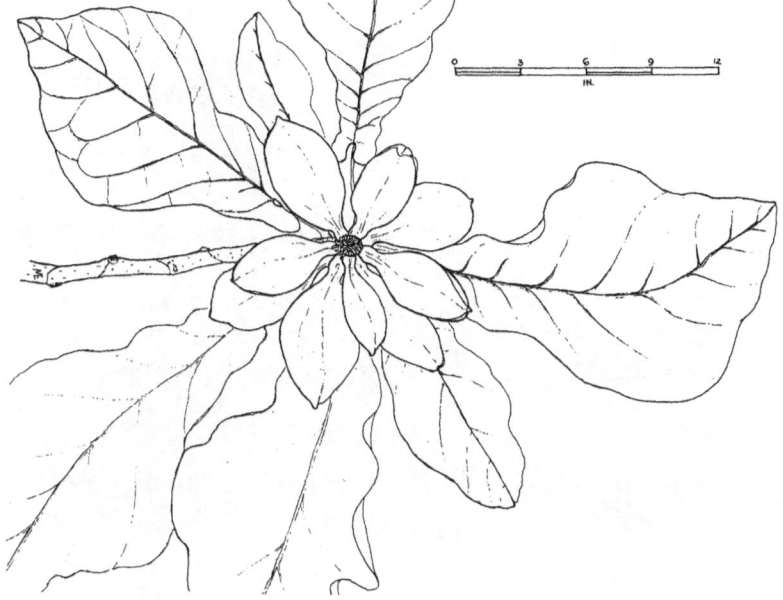

Figure 87. Magnolia pyramidata (Pyramidal Magnolia)

slits in the cone-shaped fruit, becoming bright red when they reach maturity. Yet the plant produces relatively few seeds in each fruiting structure, which bears a superficial resemblance to the closed cone of a pine tree.

————*Magnolia grandiflora,* SOUTHERN MAGNOLIA, BULL BAY (fig. 88). This is *the* magnolia that most people think about when discussing anything about the genus. Certainly it is a magnificent tree, found as a native in many of our hardwood forests, along with oaks, hickories, sweet gums, tulip trees, and others. It is planted in just about any yard or garden bigger than a postage stamp. The large, white, fragrant flowers begin blooming in May and flowers may be found on the tree for two or three months. The flowers are followed by cylindrical, tan "cones" 2–3 inches long. When the bright red seeds are mature, they project from the "cones" and hang for a while by a thin, filamentous thread. The large, evergreen leaves (6–8 inches long, 3–4 inches wide) are dark glossy green above with a tan to brown, hairy undersurface. As with nearly every plant that is

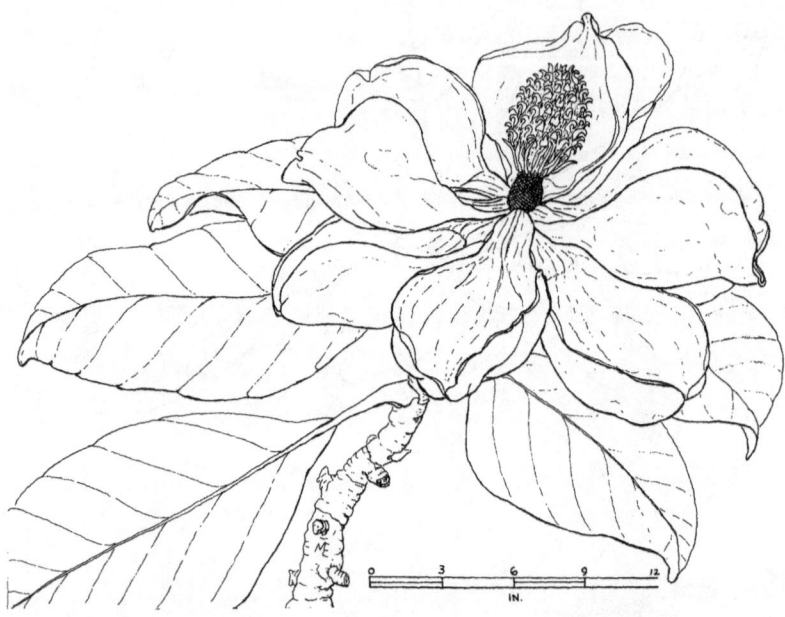

Figure 88. Magnolia grandiflora (Southern Magnolia)

highly regarded, the southern magnolia has one drawback that most of us are resigned to live with—the leaves seem to be constantly falling off, causing a mess unless they're cleaned up. They just lie there and never seem to disintegrate or break down. There are two general shapes the tree can assume when growing naturally: one is tall and cylindrical, the other is not as tall and more globose in shape. As with many cultivated plants, these trees are seldom given enough room to develop their true shapes, so that we do not often have an opportunity to see the true shape of the fully grown plant. One absolutely magnificent specimen may be seen on the south side of Lake De Funiak, on Circle Drive near Hubbard Street. Anyone who is not inspired by this beautiful tree just has no feelings at all!

————*Magnolia virginiana* var. *australis,* SILVER BAY, SWEET BAY, BAY (fig. 89). The silver bays are well known to residents of Zone 8 because just about every low spot, stream bank, or spring head has a number of these 40–60 foot tall trees whose leaves show their silvery undersides when the wind blows. We don't really need to

Figure 89. Magnolia virginiana var. *australis* (Silver Bay)

cultivate them unless we're specializing in magnolias, because they're so common and visible in their natural areas. This bay tree has rather small flowers (2–3 inches across), white and lemony-fragrant, whose structure clearly indicates this species' relationship to the magnolias. Small fruiting "cones" turn bright red when ripe and have numerous small red seeds.

Three introduced magnolia species in Zone 8.

———*Magnolia liliiflora,* LILY-FLOWERED MAGNOLIA, TULIP MAGNOLIA (fig. 90). This is the smallest of the 3 Chinese natives that have found a good home here in our zone. There are relatively few plantings of the lily-flowered magnolia, but it deserves more atten-

Figure 90. Magnolia liliiflora (Lily-flowered Magnolia)

tion. At maturity the shrub is only about 8 feet tall with many slender branches, appearing somewhat gawky in the winter when it is without leaves. The leaves are broadly elliptic, bright green, 5–6 inches long. The flowers have a bright purple outside, are rose to white inside the petals, and do not open quite as wide as flowers of most magnolias, maintaining an urn shape through their lifetime. They have no fragrance. Borne at the tip of the branches in great numbers, the flowers make a good show. Flowers appear in early spring, before the leaves. The plants do well in light shade and also seem to be happy in full sun.

―――――*Magnolia* × *soulangeana*, SAUCER MAGNOLIA (fig. 91). The parents of the saucer magnolia are two Chinese natives, *M. denudata*

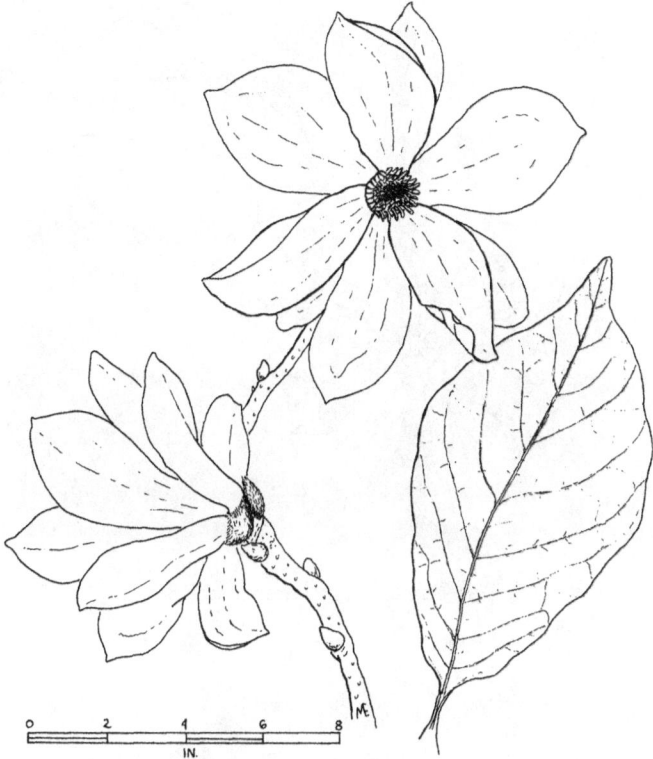

Figure 91. Magnolia × *soulangeana* (Saucer Magnolia)

and *M. liliiflora*. This is a glorious hybrid, that grows as a shrub or small tree. It flowers before its leaves appear, usually in February, but occasionally as early as late January. The bicolored, purple-to-red and white flowers are about 5 inches across and bloom in great profusion all over the tree. Though this species is cultivated as far north as Connecticut, we have never seen the plants do as well elsewhere as they do in the Deep South. After flowering, the light green foliage gives a nice, decorative flair to other plants nearby. In the winter, the handsome, light gray bark provides a point of interest.

———*Magnolia stellata,* STAR MAGNOLIA (fig. 92). This white-flowered, unscented shrub or small tree is, like all its kinfolk, well worth planting for early flowers. The many petals are more strap-shaped than spoon-shaped, as in other *Magnolia* species. This is a

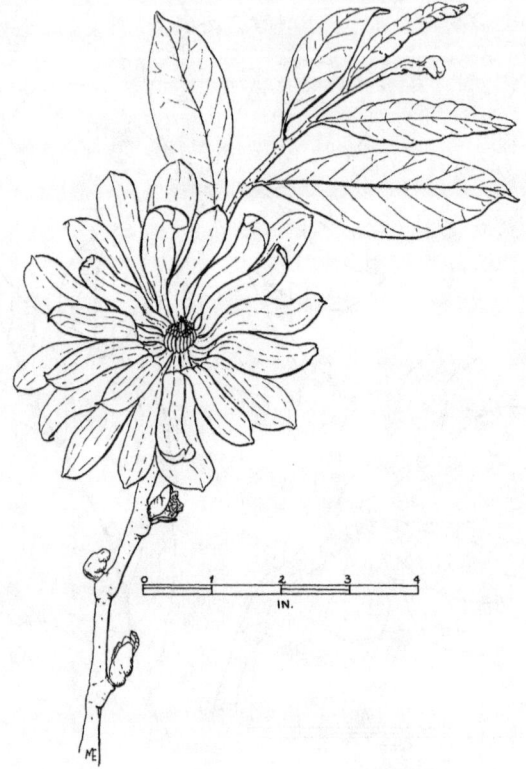

Figure 92. Magnolia stellata (Star Magnolia)

nice shrub to have as an accent plant, flowering just after most of our camellias have finished blooming and just before the azaleas begin to make their marvelous show. What a wonderful show it is when a star magnolia is placed near the early-flowering Taiwan cherry (*Prunus campanulata*)!

MAHONIA. There are about 100 species in this group and they are nearly all called Oregon grape, though they have nothing to do with the real grape genus, *Vitis*. The species are all shrubby and all have small, yellow flowers, frequently in fascicled racemes. All are characterized by compound, alternate leaves, and each leaflet has sharp spines in varying numbers. The name "Oregon grapes" probably comes from the fruits, which are blue black with a blue-gray bloom, the fruit usually smaller than a real grape.

———*Mahonia bealei*, LEATHERLEAF, HOLLY MAHONIA (fig. 93). These plants are 3 feet tall or somewhat larger, sparingly branched, the stems mostly erect. The compound leaf has from 5–9 pairs of leaflets, each leaflet with 6–12 sharp spines. The yellow flowers occur on slender branches, and each branch seemingly arises from a common point so that the whole cluster provides a sort of rounded brush shape. Oregon grape grows well in sandy soil and will withstand dry conditions for a while, but the site should be kept moist for best growth. The plant is useful as an accent plant in areas with a medium amount of shade. Since plants limit their own height there is no need for much pruning, except to remove old wood or crowded stems.

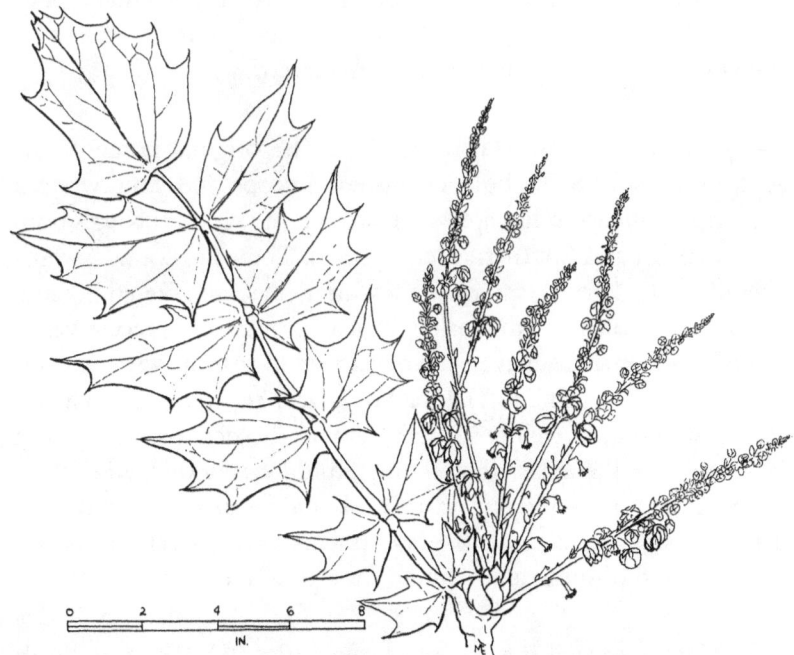

Figure 93. Mahonia bealei (Leatherleaf)

❦

MALUS. This is the genus of apples and crabapples. There are about 25 species in this genus, all of the North Temperate Zone, with many hybrids, forms and selections. *Malus* is in the rose family, and similarities among family members can best be determined by comparing the flowers and fruits of the plants. The apples and crabapples are raised both for fruit and as ornamentals, and some of our showiest shrubs and trees belong to this genus. It is beyond the intent of this work to delimit every form that can be raised here. Consult the local nursery to select the plants of this group that will fit specific needs. The experience of nursery personnel is very important as an aid to gardening and landscaping.

The fruit of plants in this genus is botanically classified as a pome, with a fleshy exterior portion surrounding a "core" where the seeds are carried. About the only real difference between the apples and crabapples is the size of the pome. The flowers are white to pinkish and may be single or double (particularly in the ornamental types). While all the apples are primarily from one species, there are several crabapple species, both native and introduced.

———*Malus sylvestris,* APPLE (fig. 94). The original home of the apple is considered to be southeastern Europe and southwestern Asia, from whence it has spread all around the world through both the Northern and Southern Hemispheres. There are countless varieties of apples known, but only a few of them are marketed because of various problems with storing, bruising, or taste. In recent years, several apple varieties have been developed in regions with climates similar to that of Zone 8. The state of Israel has produced a cultivar, 'Anna,' which does well in Zone 8; 'Dorset Golden' has come from Nassau in the Bahamas and 'Granny Smith' was developed in Australia. At least two different varieties should be planted together so that one may be assured of good pollination. Proper fertilizing, spraying, and pruning are required for a good crop.

———*Malus,* several species, CRABAPPLE (fig. 95). There are both native and introduced species of crabapples in Zone 8. As indicated above, about the only reliable characteristic differentiating the crab-

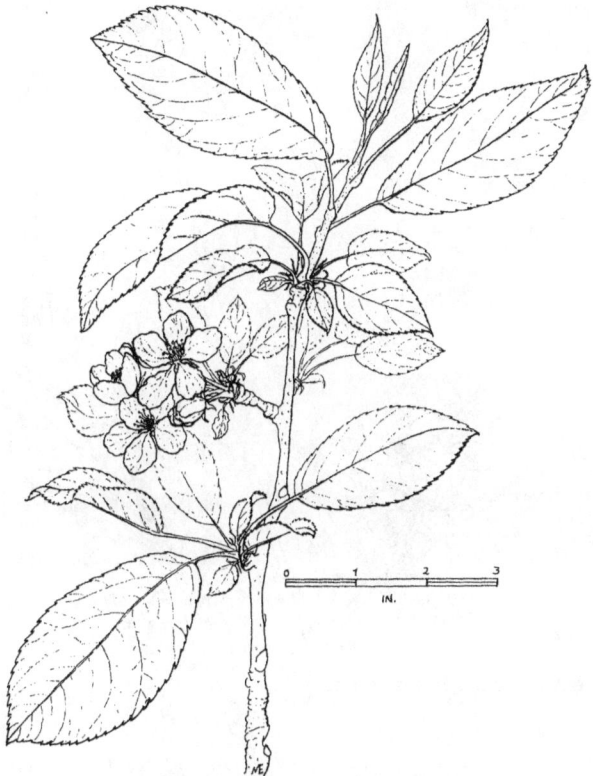

Figure 94. Malus sylvestris (Apple)

apple from the apple is the size of the fruit. The crabapples generally have a much smaller fruit. Other distinguishing characters are technical and difficult for the nonspecialist to detect or interpret properly. There are not only several crabapple species, but many hybrids between the species and many cultivated varieties. Check with your local nursery to determine the best ones for your landscaping needs. Crabapples are well worth the effort because they, like a few other well-known cultivated plants, are the real harbingers of spring. In addition to being raised for their floral beauty, some crabapples are harvested for their fruits and made into jellies or jams.

MALVAVISCUS. This is a small genus of three species in the same family, Malvaceae, as cotton and hibiscus. The species described

126 ❧ *Malvaviscus*

Figure 95. Malus species (Crabapple)

here is usually easy to propagate from stem cuttings of new wood kept in moist condition (mist is preferred) in a medium of sand, peatmoss, and perlite. These plants are mostly from the tropics, but this species does well here in Zone 8.

——————*Malvaviscus arboreus* var. *mexicanus,* TURK'S CAP HIBISCUS (fig. 96). This is a low shrub, growing up to about 4 feet tall and spreading by runners, but not objectionably. This variety will form a small clump, and plants collectively make a very good show with the bright red flowers occurring along the upper parts of the stem. The flowers are particularly welcome, appearing from September up to the first frost. Furthermore, the plants are quite hardy; ours have survived three rather severe winters when many other plants,

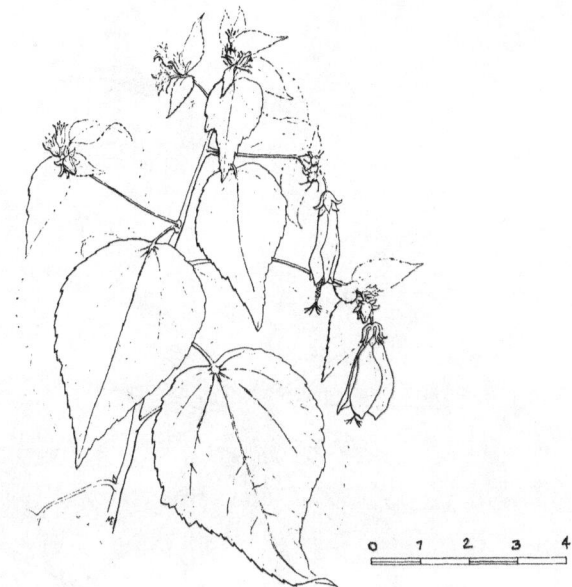

Figure 96. Malvaviscus arboreus var. *mexicanus* (Turk's Cap Hibiscus)

thought to be quite hardy, have been killed back. No special care is needed past the early stages of establishment.

MANIHOT. This is a genus of eighty-seven species, all from tropical America, a group in the spurge family. Other familiar genera in the spurge family are the poinsettia, snow-on-the-mountain, and castor oil plants. Only one or two species of the genus *Manihot* will grow outside the tropics, one of which is *Manihot grahamii,* or wild cassava. This is not to be confused with the tropical cassava, plants cultivated for tapioca, which are members of the same genus.

———*Manihot grahamii,* WILD CASSAVA (fig. 97). This species came to the United States as a stowaway aboard a ship that must have taken on cargo somewhere on the La Plata River in South America, then put ashore at the port of New Orleans. The first records of wild cassava in this country come from the areas around New Or-

Manihot

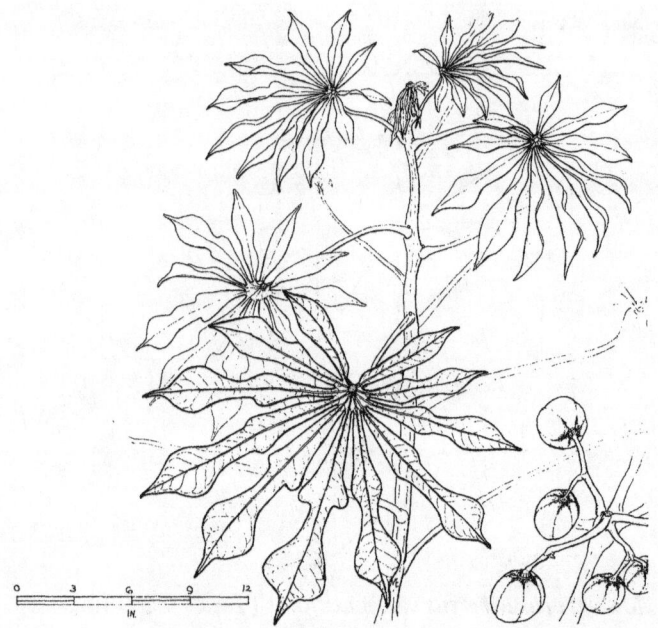

Figure 97. Manihot grahamii (Wild Cassava)

leans, but the plants have spread rapidly to the rest of Zone 8. Most of the plants are shrubs, but they will, if not killed back by frost, grow to be small trees with a single trunk. If grown in the open, the crown of the tree is a handsome umbrella shape. The bark is a smooth reddish brown. The leaves are the most striking part of the plant, with eight to ten strap-shaped lobes radiating from a common basal area on long, gracefully curving petioles (leaf stalks). The flowers are small and yellow-green, not very conspicuous organs, and the fruits are round and dry at maturity. The seeds are scattered when the dry fruit shatters and seem to germinate readily. Birds are probably responsible for the rather rapid distribution of the plants throughout Zone 8. If one is interested in raising plants that are different from other, better-known cultivated plants, this will make a good candidate. Wild cassava should be grown in protected, sunny areas, in almost any soil. It actually thrives in drier areas, and needs little care.

MELIA. This genus is in the same family as the true mahogany, Honduras mahogany. There are about ten species in *Melia*, but the one below is the only one in cultivation.

————*Melia azedarach*, CHINABERRY (fig. 98). This is one of the South's most common shade trees, having become naturalized in many areas. The chinaberry was introduced from Asia in the eighteenth century. It is a deciduous tree with a (usually) short trunk and spreading branches, making a nicely rounded or umbrella-

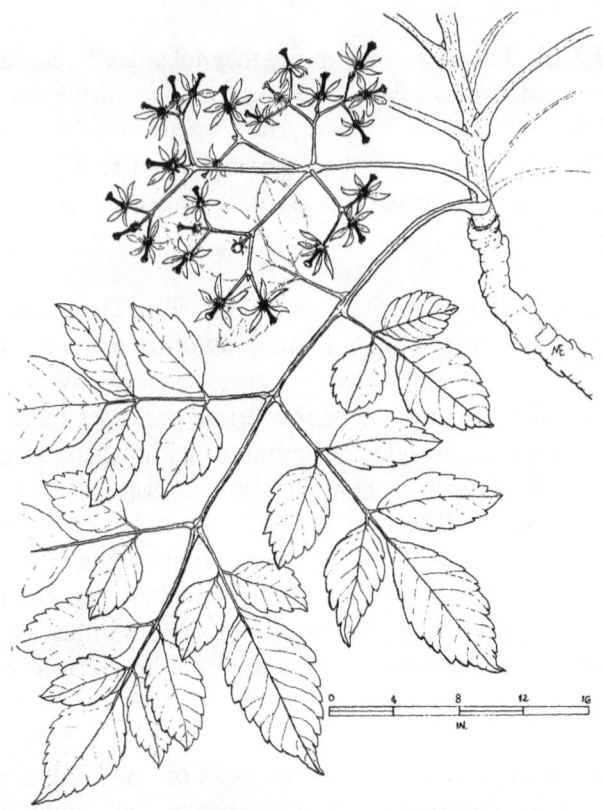

Figure 98. Melia azedarach (Chinaberry)

shaped tree. The wood, unlike that of its cousin Honduras mahogany, is very light and weak, not good for furniture or construction. In the summer, the trees produce many small, fragrant, lavender flowers in panicles, and these are very attractive to honey bees. Unfortunately, the honey derived solely from this tree has something of a bitter flavor. The fruits are about ½ inch across, yellowish when ripe, and bitter tasting. When DJR was a boy, the green, hard, round fruits were used as ammunition in homemade pop guns. It is a wonder we didn't get sick, because the fruits are said to contain an unknown poisonous principle, and kids do have a way of putting such things in their mouths!

❦

MICHELIA. This is a genus in the magnolia family, coming from temperate and tropical Asia. They are trees or shrubs, evergreen, and the flowers do have a rather close resemblance to magnolias, but sufficiently different to be separated from that genus. Most members of this group have pleasant fragrances, which does have some resemblance to magnolias.

─────*Michelia figo,* BANANA SHRUB (fig. 99). The banana shrub has long been a favorite in De Funiak and other places for its pleasant aroma, which is indeed reminiscent of bananas. We have heard that, before the advent of reasonably priced perfumes and colognes, some young ladies carried handkerchieves with banana shrub flowers. Perhaps so. The shrub grows to about 15 feet tall and is a pleasing globular shape. The flowers are 1–1½ inches across with numerous petals, yellow, with a greenish tinge along the margins. Propagation is carried out largely from cuttings, but the cuttings must be taken from medium-hardened wood and must be kept in high humidity and warm temperatures if they are to strike roots.

❦

MORUS. This genus of about ten species of deciduous trees has a widespread distribution pattern, with one or a few species found in North and South America, Africa, and Asia. All species are called mulberries. Perhaps the best known species is the silkworm mul-

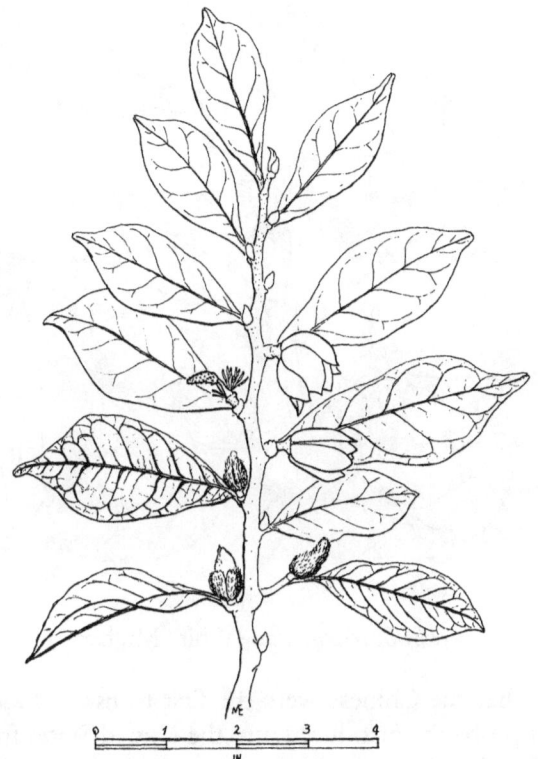

Figure 99. Michelia figo (Banana Shrub)

berry, native to China. There is one native Florida species, the red mulberry, but this one is less likely to be seen in yards or gardens than is the white mulberry, discussed below. Since all the species produce juicy fruits that somewhat resemble blackberries, they are good candidates to be planted, especially for those of us who wish to provide food for birds. All the species have separate male and female flowers, both borne on the same tree.

———*Morus alba*, WHITE MULBERRY (fig. 100). There are several cultivars of this tree. It has glossy green, ovate leaves that are coarsely serrate on the margins. The fruits can be white, pinkish, or blackish-purple. Those in Zone 8 have blackish-purple fruits that can be messy if they fall on your roof or car. This tree, which has

Figure 100. Morus alba (White Mulberry)

one variant that the Chinese were the first to use as food for silkworms, was probably introduced into the United States from China to establish a silk industry in the Deep South. Evidently the industry did not get going, but the trees have become well adapted to this climate and have become naturalized. If you wish to plant a mulberry, it will be easy to dig up a root sprout and transplant it—but don't place it too close to the house!

MYRICA. This is a genus with several native American species, one of which is the bayberry. A candle wax derived from the fruits of these plants has a pleasant odor when the candles are burned. All the species are generally similar in appearance and all of them have leaves that are pleasantly aromatic.

———*Myrica cerifera,* WAX BAYBERRY, WAX MYRTLE (fig. 101). The berries produce a waxy covering and are generally sort of bluish black in the late fall. This evergreen species, whose flowers are in-

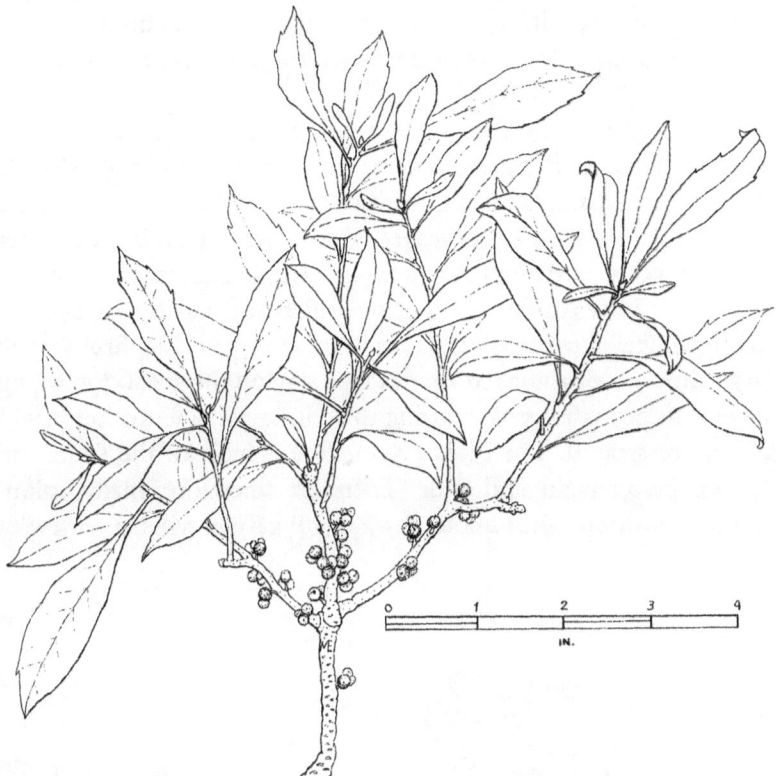

Figure 101. Myrica cerifera (Wax Bayberry)

conspicuous, is so common that it was not thought of as a decorative or ornamental plant until some enterprising landscape gardener discovered the value of this large bush or small tree. It grows vigorously in almost any circumstance, from bogs to high, dry pineland. When you crush the leaves in your hand, you get a rather pleasant odor, but it fades away rapidly. Use this plant as a background, in hedges, or as an evergreen to grow in that "difficult" area of your yard.

NANDINA. This evergreen shrub, a native from India to east Asia, has long been one of the more useful shrubs planted in Zone 8. There is only one species, which apparently is known only as a cultivated plant even in its country of origin, for the authority that named it called it *domestica,* a term used to indicate cultivation.

———*Nandina domestica,* HEAVENLY BAMBOO, NANDINA (fig. 102). This is an evergreen shrub growing 6–7 feet tall. Usually there are many stems arising from the base and as the stems grow older they shed their lower leaves. The leaves, which last well after they are picked, are frequently used in floral arrangements. They are compound leaves divided into many leaflets and may turn a dark red in the winter. The major attraction of these dependable shrubs is the large clusters of bright red berries that stay on the shrub for a long time, thus providing color during the winter when few other bright colors are around. The birds eventually eat most of the fruits and scatter the seeds far and wide. There are, therefore, many volunteer plants around. Just about any Zone 8 garden will have a few of

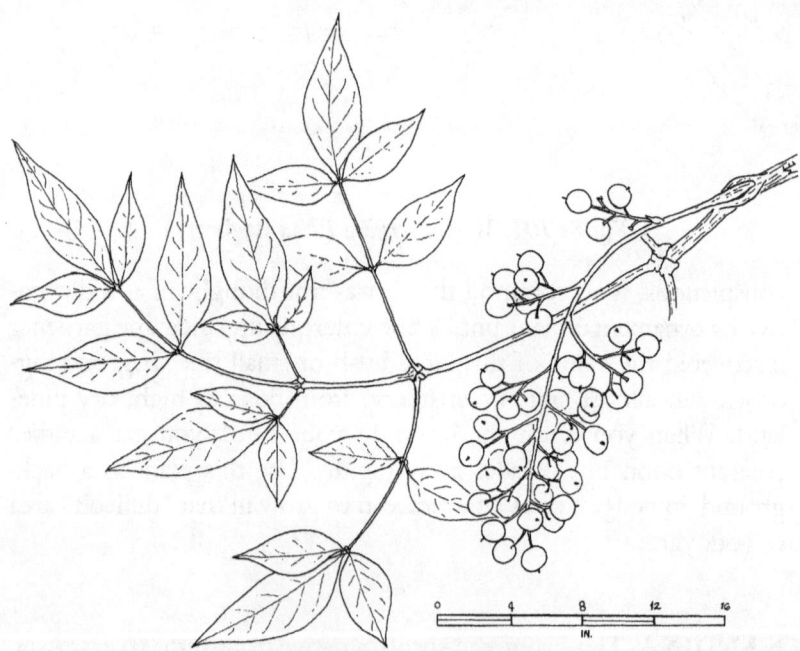

Figure 102. Nandina domestica (Heavenly Bamboo)

these plants springing up in unwanted areas. They also transplant easily.

❀

NERIUM. The one species of this genus, oleander, is a native of the Mediterranean region and has long been a shrub of choice in southern gardens. The slender, erect stems may reach 8–10 feet, and several arise in a clump. The leaves are evergreen, and usually there are three leaves arising in whorls at each node or joint along the stem. The flowers are 1–2 inches across and have a very interesting, attractive form. They may be either white, red, pink, or, in some selected varieties, variegated. The plants require practically no attention and will withstand drought and open sun.

————*Nerium oleander* (fig. 103). The species name of this plant, *oleander* is its common name as well. There is a legend as to how this plant got its name. As the story goes, there was in ancient times a beautiful young Greek damsel named Leander. She was in love with someone her family thought to be beneath them so they took her far away. The damsel's lover naturally began looking for her, but was about to despair of ever finding her. One day in his distraught condition, he was sitting down near some shrubs bewailing the loss of his loved one: "Oh, Leander where are you?" Miraculously, there she was, just behind the bush which has ever since borne that name! We don't suppose it hurts this charming story to know that the plants are poisonous in all parts. If young children are around, it might be good to warn them against chewing anything from this plant. Additionally, it is said that inhaled smoke from burning oleander leaves is also poisonous.

❀

NYSSA. This is a small genus of deciduous tall shrubs or medium height trees. The 8–10 species (there is a difference of botanical opinion about the number of species) are found either in North America or eastern Asia. The name *Nyssa* is derived from a Greek

Figure 103. Nerium oleander (Oleander)

word for nymph, and alludes to the habitat of the species, either in swamps or low, moist areas.

———*Nyssa sylvatica,* SOUR GUM, TUPELO. The sour gum tree grows to about 50 or 60 feet, or, rarely, reaches a maximum of 90 feet. This native tree of the eastern United States is deciduous and its main major horticultural value is in the foliage—glossy-green in summer and bright orange to red in fall. Like the red maple, *Acer rubrum,* the natural habitat of the sour gum is in very wet, low-lying areas, either as a companion tree to the red maple or the silver bay tree (*Magnolia virginiana* var. *australis*). It will, however, grow quite well in upland areas, requiring very little more care than frequent watering just after being planted. We have never tried to transplant one from the wild, but have found good plants in some nurseries, particularly those that specialize in native plant materials. The sour gum and a closely related species, *Nyssa aquatica,* are both referred to as tupelos and both are the source of a very fine honey, also called tupelo. Almost everyone in the Deep South knows about tupelo honey, and that it is one of the finest flavors of all honeys. It is said that bee-keepers put their hives on barges, and in the early spring start at the lower end of the rivers and move slowly up the rivers as the season (and flowering) progresses.

OPUNTIA. There are many species of the prickly pear cactus, perhaps more than 300, but most of them are found in much drier regions than Zone 8. However, when Zone 8 gets dry weather in the summer it is almost like a desert in sandy areas where the water drains out very rapidly. All members of the cactus family except one are native to the warmer regions of the western hemisphere. The one exception is a genus from West Africa, *Rhipsalis.*

————*Opuntia ficus-indica,* INDIAN FIG CACTUS, PRICKLY PEAR (fig. 104). The species name gives the false impression that this prickly pear is like the fig, *Ficus,* and that it is from India, which it isn't. There are no cacti native to that country and this one no doubt came from somewhere in North America, but no one knows where. It is cultivated in Zone 8. This is a much larger plant than several native species that are found in Deep South piney woods (which no one bothers to cultivate). The broad, leaflike pads (usually called "joints") are actually stems, and what serves for leaves in this group are very small, scaly structures around the small clusters of tiny spines (or glochids). There are relatively few long spines on the pads, and in some selections there may be no spines at all. The flowers, usually arising along the upper edges of the joints, are yellow green, and the fleshy fruits are purple or reddish with green skin. You can propagate these cacti by taking a joint and burying it up to half its length in the soil; no other treatment is necessary. The plants, if placed at a corner, can prevent people taking shortcuts across private property!

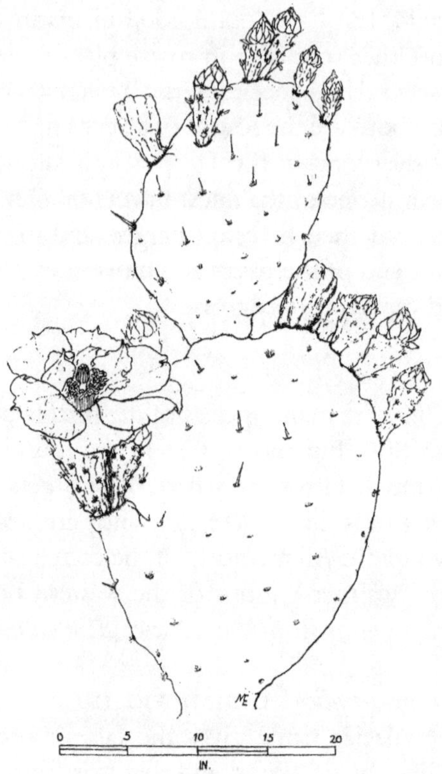

Figure 104. Opuntia ficus-indica (Indian Fig Cactus)

OSMANTHUS. This is a small genus of 30 to 40 species of evergreen shrubs and trees, native to east Asia but also found in North America and a few other localities around the world. The genus is in the olive family.

———*Osmanthus americanus,* DEVILWOOD, AMERICAN OLIVE, WILD OLIVE (fig. 105). This is a tree native to the southeastern United States, growing to about 45 feet tall. The evergreen leaves are elliptic, smooth-edged, and glossy and make a good show in any season. The fragrant flowers are a creamy or yellowish white, rather

Figure 105. Osmanthus americana (Devilwood) After Britton, *North American Trees*, 814, 1908.

small, in clusters in the axils of the leaves. The fruit, like that of other members of the genus, resembles a small olive.

————*Osmanthus fragrans,* TEA OLIVE, FRAGRANT OLIVE (fig. 106). As the species name implies, the many small, white flowers are very fragrant. It is also called tea olive because, in China, the flowers are used to flavor tea. The plants can grow to 30 feet tall, but mostly one sees them as medium-sized shrubs. The leaves are a nice green, elliptic and fine-toothed. Cuttings made in the late summer from half-ripe wood and placed under high humidity will root. Seeds take two years to germinate, and often may not be produced on plants cultivated in Zone 8.

140 ❦ *Oxydendron*

Figure 106. Osmanthus fragrans (Tea Olive)

OXYDENDRON. This genus, known as sourwood, has only a single species and is found in the eastern United States, its range just barely extending into north Florida.

———*Oxydendron arboreum,* SOURWOOD (fig. 107). This is a native in rich, mixed hardwood and softwood forests of most of the eastern United States. It is a tree growing to 25 feet tall or higher. It is a member of the same family as blueberries and mountain laurel. The 3–4 inch, slender, lance-shaped leaves, fine-toothed on the margins, glossy green in summer, turn scarlet in the fall. The flowers

Figure 107. Oxydendron arboreum (Sourwood)

are borne on drooping stalks, which remain to hold the fruits in late summer. Sourwood is seldom planted in our yards but can be used successfully as a background tree where a little fall color is needed.

❊

PARKINSONIA. This is a small genus of two species in the bean family Leguminosae. One species is from America, the other from Asia. The plant cultivated in Zone 8 is the American species, occurring mostly in Mexico, in drier regions. One might assume that this would exclude it from successful growth in Zone 8, but this region's deep sands with their rapid drainage almost act like a desert. It is only the frequency of rain that makes the difference.

Parkinsonia

———*Parkinsonia aculeata*, JERUSALEM THORN, MEXICAN PALO VERDE (fig. 108). The fun of common names is exhibited here: this plant has no real connection with Jerusalem, but the species does have many stout thorns and one might think the plants could have been used as the crown of thorns worn by Jesus. However, there was no need to import thorns to Israel since they are well supplied with many sticky plants of their own. Furthermore, it is unlikely that the plant under discussion, which grows in America, would have been known in Israel in those days! The Jerusalem thorn is a small tree growing to perhaps 20 feet tall. It has large, feathery, light green leaves (which, no doubt leads to the second common name Mexican palo verde). There are stout thorns, some more than an

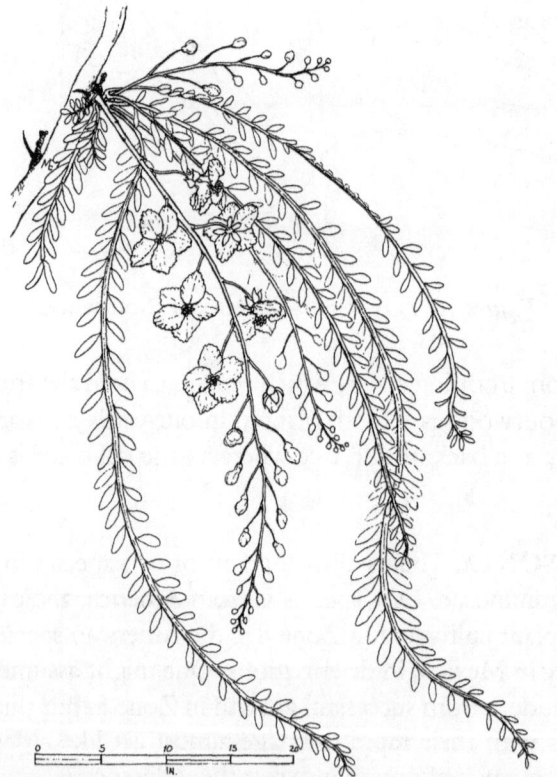

Figure 108. Parkinsonia aculeata (Jerusalem Thorn)

inch long, on the trunks. Dr. Fannie Fern Davis, a well-known botanist in Zone 8, said that she trimmed off the thorns to make the plant less offensive. The yellow flowers are about 1 inch across and are borne in racemes in the same arrangement as those of the *Wisteria*. Again, according to Dr. Davis, the plants grow very easily from seed. However, the species may be cold-sensitive.

PARTHENOCISSUS. The common name given to all 15 species of this genus is woodbine. All are vines and are native either in Asia or in America. The plants belong to the same family as the grapes, Vitaceae. The vines hold on to whatever they are climbing with small, cup-shaped disks at the tips of tendrils. The leaves of the species are all deciduous and in the fall may turn a brilliant red before they drop.

————*Parthenocissus quinquefolia*, VIRGINIA CREEPER, WOODBINE, AMERICAN IVY (fig. 109). This native American species is the one most frequently used horticulturally. Some may recall the television show "Halls of Ivy," about life on a college campus. The ivy climbing up the walls of the buildings on many campuses is as frequently the Virginia creeper as it is English or "true" ivy. There has been some controversy as to whether or not the vines climbing brick walls cause the mortar and bricks to deteriorate. It has also been claimed that ivy on the walls will help keep buildings cooler in summer by shading the walls from the sun, but neither of these claims has been properly evaluated. Our own preference is to use ivy on walls—it does enhance buildings, making them more interesting structures both in summer and winter. The vines can also be grown on trellises or arbors. Propagation is by seed, rooted cuttings, or layering (see description of layering under *Deutzia*).

PAULOWNIA. A genus of only about six species native to China. All are trees with deciduous leaves. These grow mostly in the warmer parts of the United States, though some are found along coastal areas as far north as New England. In some cases the trees

144 ❦ *Paulownia*

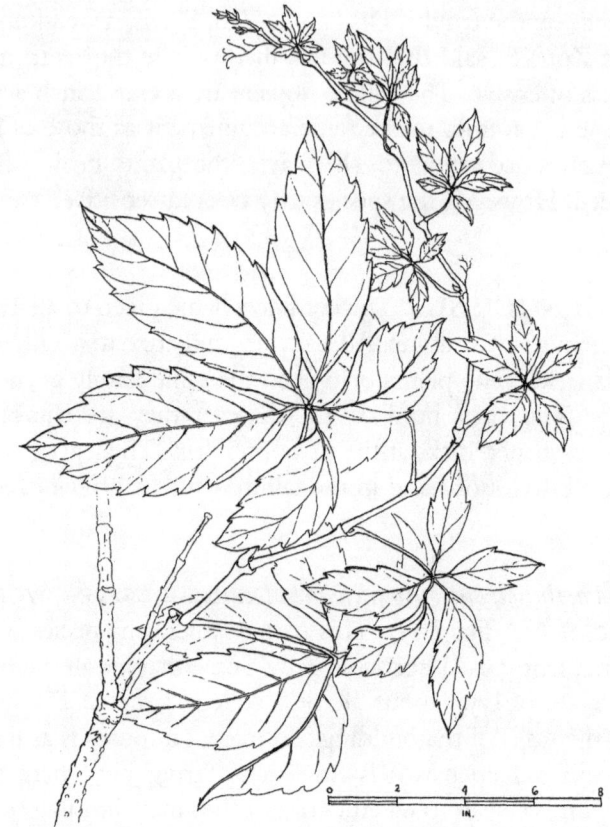

Figure 109. Parthenocissus quinquefolia (Virginia Creeper)

have escaped cultivation and grow prolifically in sheltered habitats. The most common species, *P. tomentosa* is found in moister regions of the eastern United States, in parts of California, and in the Pacific Northwest.

———*Paulownia tomentosa,* PRINCESS TREE (fig. 110). One representative of this tree occurs in De Funiak Springs. It is a medium-sized tree, 30–60 feet tall, with large leaves reminiscent of the catalpa, to which the princess tree is related. The fragrant flowers are 1–2 inches long, irregularly tubular in shape, and pale violet to purple shade. The flowers are clustered in a large panicle and make

Philadelphus ❁ 145

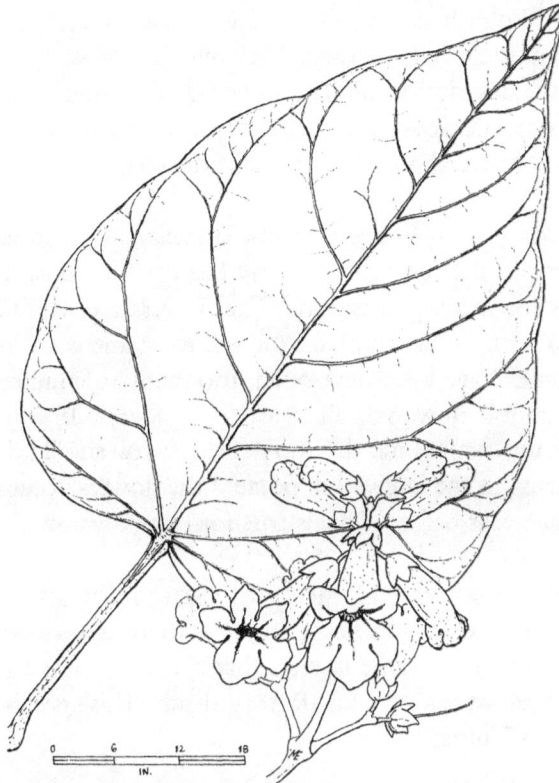

Figure 110. Paulownia tomentosa (Princess Tree). After Britton and Brown, *Illustrated Flora of the Northeastern States and Canada,* vol. 3, 189, 1913.

very striking blooms; these are followed by large numbers of dry fruiting structures that resemble clusters of grapes. They can be used in dry arrangements. This is an interesting tree, but one that is easily replaced by more well-adapted cultigens.

PHILADELPHUS. There are about 65 species of shrubs in this genus. You might ask, why do you say "about"? Simply, there is no definitive study of the group, or because there are several and the authorities disagree. So we take the middle road and say "about" this number. All the shrubs are called mock orange, but some are also called "English dogwood." There are two species in Zone 8, one fragrant, the other odorless. They may be propagated by seeds, layering (see description under *Deutzia*), or cuttings, the latter in summer from soft wood, but the best method for most gardeners is to buy good material guaranteed by the nursery.

―――――*Philadelphus coronarius,* MOCK ORANGE. This sweet-scented shrub is one of the old favorites, and has been a pleasant addition to gardens for many generations. The flowers are white and have four petals with a cluster of bright yellow stamens in the center. Many varieties have been developed, too many to enumerate here; some are double-flowered, all seem to be sweet. If you want to prune back the shrubs, and they do get leggy, you should do so right after flowering in the late spring because the flowers occur on wood produced the year before. This is true for many shrubs.

―――――*Philadelphus inodorus,* MOCK ORANGE, ENGLISH DOGWOOD (fig. 111). As the species name implies, there is no odor to this shrub. It is about the same in size, shape, and color as the scented one and therefore is a very satisfactory shrub. This species is native to the eastern United States.

❦

PHOTINIA. This is a small group of species, some of which are deciduous, others evergreen. Two evergreen species and a hybrid are of interest, all from Japan.

―――――*Photinia* × *fraseri (P. serrulata* × *P. glabra),* RED TOP, RED TIP (fig. 112). The plants are used mostly in hedges and the young leaves at the tips of the branches are a beautiful, translucent, bright red. With proper pruning, the young leaves can be kept coming through much of the summer. These plants are sometimes grown in the same hedge with a gray-green leaved shrub, *Elaeagnus pungens,* the silver thorn, and the combination makes a most striking and handsome spectacle. If the plants are left to grow as individuals, they will produce flower clusters with many small, white flowers that make a good show, but such plants will not have as good a

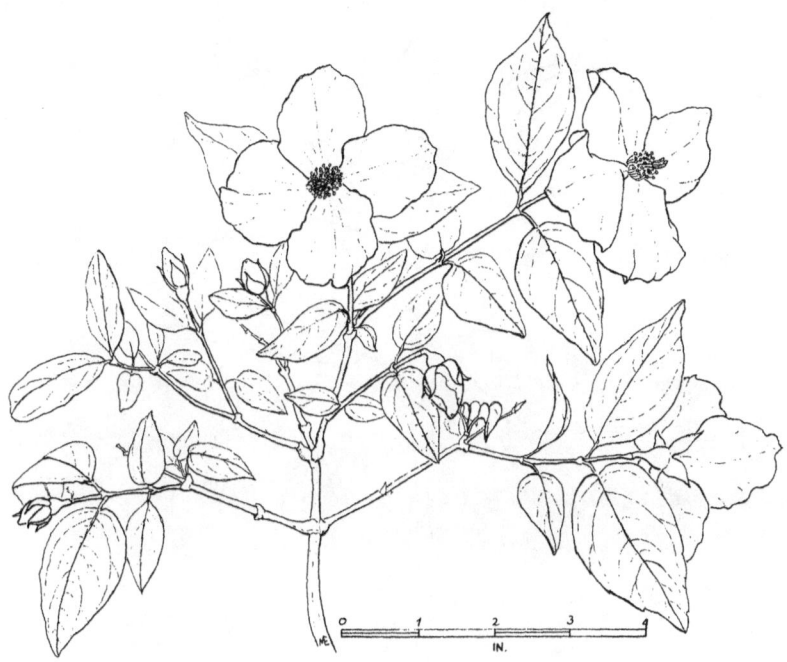

Figure 111. Philadelphus inodorus (Mock Orange)

show of leaves as those that are kept pruned. This species has been introduced to Zone 8 rather recently, or at least came to be very popular recently. We do not recall seeing any of these plants here before World War II. The hybrid species was produced at Ollie Fraser's nursery in Montgomery, Alabama.

―――*Photinia serrulata* (fig. 113). We do not know a common name for this species. The foliage, while attractive, does not have the bright red younger leaves of the hybrid above, so the hybrid must have derived the red tips from the second species of the cross, *Photinia glabra*. The leaves are oblong, 4–5 inches long, and the margins have very small serrations (that is, they are serrulate). This species is either a tall shrub or a small tree, up to about 20 feet tall (though they are not commonly seen taller than 15 feet). If left to grow naturally, the plant will assume a rounded shape and produce numerous flower clusters (inflorescences) with many small white

148   *Phyllostachys*

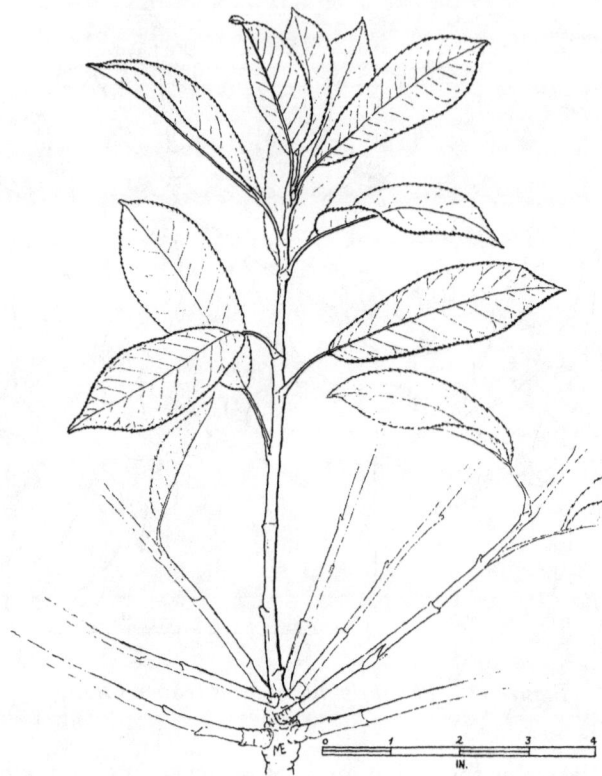

Figure 112. Photonia × fraseri (Red Top)

flowers, the whole inflorescence being as large as 4–5 inches across. Flowering usually occurs in April. The plants may be used to fill in a corner or angle in a house, preferably in full sun. They should be allowed 10–15 feet diameter to develop normally.

PHYLLOSTACHYS. All of the 30 or more species of this genus, natives of east Asia, are called bamboos. The bamboos, or "tree-grasses," are members of the grass family, which is probably the most important family of flowering plants as far as humans are concerned. There are a few other woody genera of the grass family, one of which, the giant reed *Arundo donax,* was mentioned earlier. But some of the bamboo species grow much larger and taller—as tall as

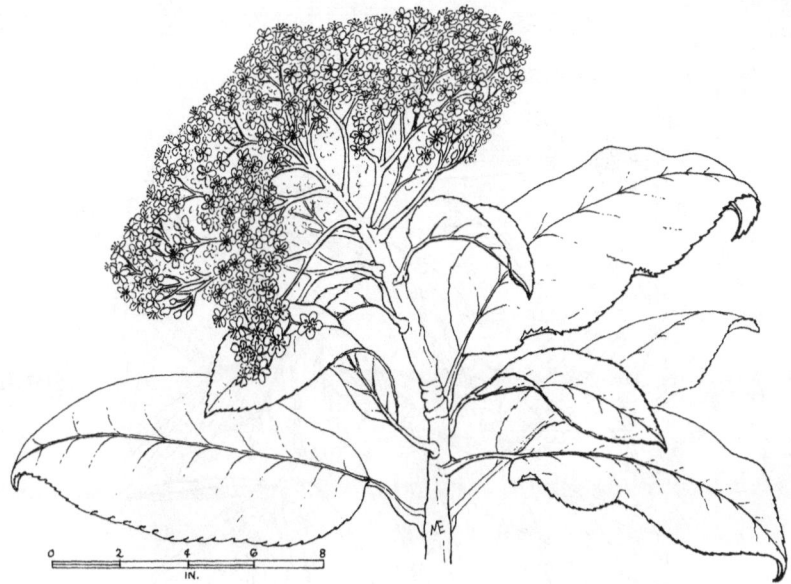

Figure 113. Photinia serrulata (no common name)

100 feet—than any other member of the family. There are multiple uses of the bamboos, particularly in southeast Asia and adjoining islands. The very young shoots of several species are frequently an ingredient of Chinese foods, and other species are used in construction of buildings, pipe lines, roofing, and paper. About the only item made of the species described here is fishing poles, but the young shoots may be eaten.

————*Phyllostachys aurea,* FISHPOLE BAMBOO, BAMBOO (fig. 114). These plants grow in ever-widening clumps, spreading by vigorous underground stems (technically known as rhizomes). They may become a nuisance because of their vigorous growth but, with some modern-day weed killers, they can be kept in bounds. The jointed, hollow stems are usually green when young, turning yellow or golden at maturity. The stems may be over 1 inch in diameter and are very strong for their weight. In addition to fishing poles, they make excellent staking material for weak-stemmed plants, such as some of the large varieties of gladiolus. As mentioned above, the

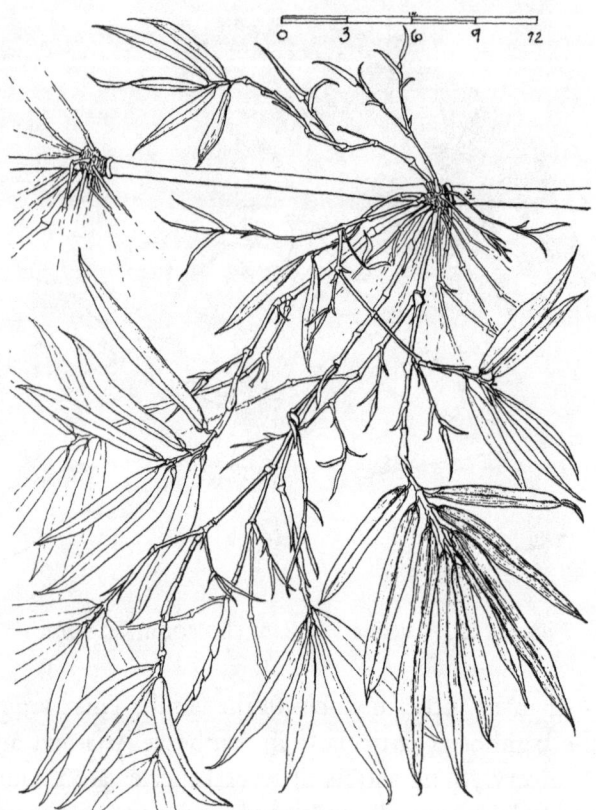

Figure 114. Phyllostachys aurea (Fishpole Bamboo)

stems may be eaten when they are very young. These plants seldom flower, but when they do flower and produce seed, they usually die completely.

❈

PINUS. The pines are so much a part of the Zone 8 landscape that we take them for granted. We don't consider them as cultivated plants, if we consider them at all. However, as we learned, if you move away from the Deep South to areas where there aren't any pines, you begin to miss them. When you return, your spirit jumps with joy when you first approach "home" country with its pines. Most of us just accept that pines are pines, and without considering there are at least five, and maybe six different pine species that are native to Zone 8. Of course, the pines were one of the major reasons for settlement of this part of the country—the native forests were prime sources of some of the best timber anywhere in the United States. In addition, Zone 8 pines produce a "pitch" or resin that exudes from the trunk when it is slashed through the bark. This resin was harvested in quantities sufficient to establish another industry, the turpentine industry (see description under *Pinus elliottii*).

In the last 50 years or so, the pine tree has provided the country with one of its most important renewable resources as the basis for the paper industry. Most west Florida counties south of the good soil belts in the northern sectors are not suited for regular agriculture, but these regions can be made profitable as pine tree pulp plantations. Southern pines grow much more rapidly than their northern relatives and can produce a trunk large enough in fifteen years or so that they can be harvested for pulp wood. Altogether there are about ninety species of pines growing in many localities around the world, largely in the northern hemisphere. Many are tall, stately trees, evergreen with needle leaves; needles of some species are very short while those of others are well over a foot long. Some species are shrubby and others just hug the ground. Their shapes may be conical or cylindrical, but frequently, as these trees approach maturity, their crowns develop an umbrella shape. You can usually tell the really mature specimens from those that are not yet mature by their shape: the older the tree, the more umbrella-shaped it becomes.

There are many ways to distinguish between the different species of pine, but the easiest is to observe the characteristics of needles and cones. The needles are usually in fascicles or bundles of two, three, or five; the length of the needles is also a characteristic of a species. The cones vary in size and shape and in the thickness of individual cone scales. The point at the end of the cone scales is sharp on most species but absent on others. There are many other distinguishing features, but these are the ones most easily seen.

All of the species described below except *Pinus strobus* and *P. virginiana* are natives. Anyone interested in the proper culture of the pines (or of several hardwood species as well), where to obtain

152 ❦ *Pinus*

plants, and almost any other problem with these trees, is advised to consult the county forester, whose telephone number is usually listed under the State offices in the phone book.

———*Pinus clausa*, SAND PINE, SCRUB PINE (fig. 115). Two varieties of the sand pine are recognized: the Choctawhatchee and the Ocala. Foresters generally recommend the first of these two for planting in Zone 8. The sand pine grows, as indicated, in the very deep, almost sterile sandy soil found in many parts of the region. The Florida Forestry Division recommends this pine for pulp plantations in these regions. These trees have very short needles, in bunches of two, about 3 inches long. The cones are conical when closed and ovate when open, with a short, stout spine on each scale face. The cones are very small, 2–3 inches long. The sand pine is a

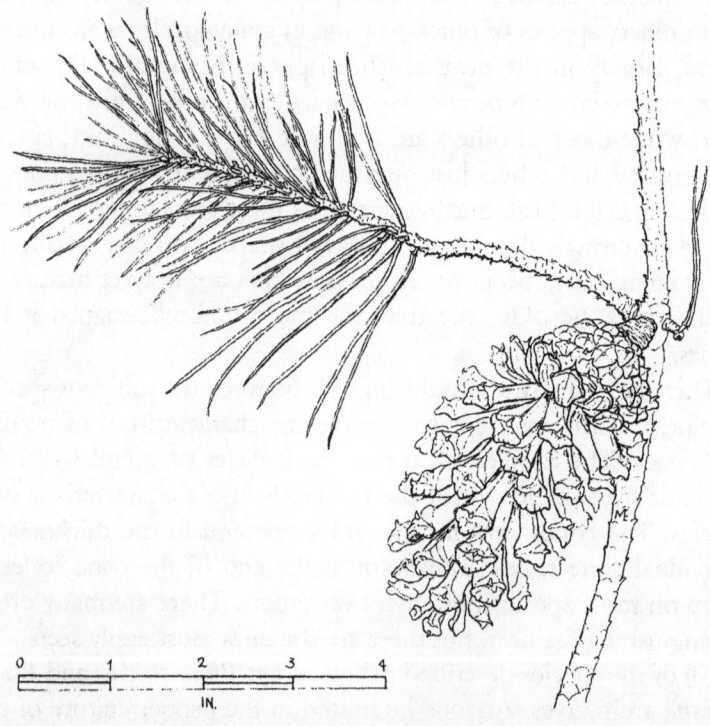

Figure 115. Pinus clausa (Sand Pine)

very good species for outdoor Christmas trees if managed properly. The needles are placed to make a more closed appearance on the branches than other species where the branches are set more widely apart.

———*Pinus echinata,* SHORTLEAF PINE, SHORTLEAVED YELLOW PINE (fig. 116). This pine, slash pine, and the loblolly pine all grow in more or less the same localities, at least as planted—their native distributions might have been quite distinct. The shortleaf has needles in fascicles of 2, or sometimes 3 that are about 3–5 inches long. The cones are small, 1½-2½ inches long, conical when closed, oval when open. Their spines are short and weak, sometimes absent on the scale-face.

Figure 116. Pinus echinata (Shortleaf Pine)

———*Pinus elliottii,* SLASH PINE, SWAMP PINE (fig. 117). The slash pine is probably the pine one sees most frequently these days because the Forest Service has recommended it over any other for pulpwood production. Its needles are in bunches of two or three, 7–12 inches long. The cones are much larger than those of the species described above, 3–4 inches long, narrowly ovate when closed, broadly ovate when open, with small, straight or recurved spines on each scale-face. Although its greatest use by far is for pulpwood, its dried needles are used in the weaving of baskets. Lumber is also an important product of this tree, both sawed boards and roundwood poles. The reason for the common name "slash pine" is probably the fact that it produces so-called naval stores or resin in abundance

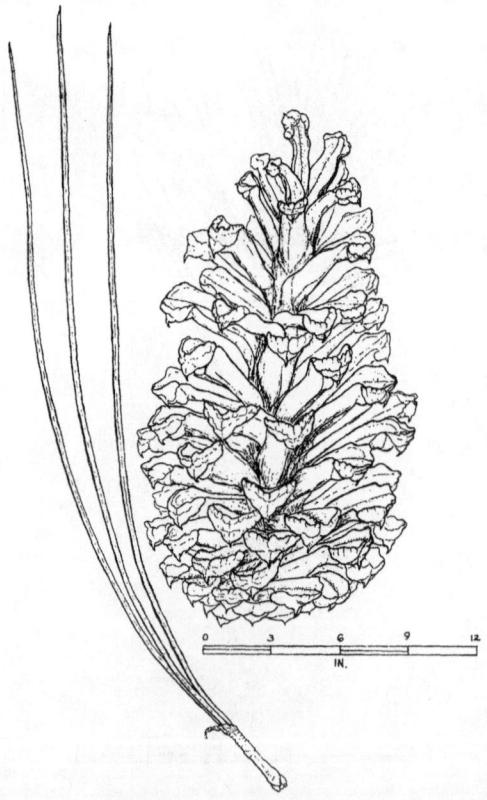

Figure 117. Pinus elliottii (Slash Pine)

and, in earlier times, men with specially shaped knives slashed the bark to cause the resin to run down the trunks into buckets.

———*Pinus glabra,* SPRUCE PINE (fig. 118). This pine is being offered by the Forestry Division for reforestation in Zone 8, under certain conditions. If you want a thousand plants or more, you can get them by ordering through your county forester. Few gardeners want this many plants, but sometimes a friend can spare a few out of his order. This species is a short-needled pine, its needles about 3 inches long, in bundles of two or three. The needles are very flexible. The cones are quite small, conical when closed, ovate to rounded when open, about 1–2 inches long. This species could make a very nice Christmas tree if properly pruned.

Figure 118. Pinus glabra (Spruce Pine)

———*Pinus palustris,* LONGLEAF PINE, YELLOW PINE, SOUTHERN PINE (fig. 119). This is the most magnificent pine of all in Zone 8, the tallest with the longest needles (in bunches of three, 8–18 inches long). In the wild it is very slow growing. It is, however, being recommended for reforestation under certain special conditions. Consult the local forester for his recommendations. The longleaf pine used to cover much of the region, but indiscriminate cutting without replanting nearly drove it to extinction. It is now coming back because of the influence of ecologists and environmentalists. In the "grass" stage of its growth cycle, the longleaf, for one to three years, is a short-stemmed plant with one terminal growth of

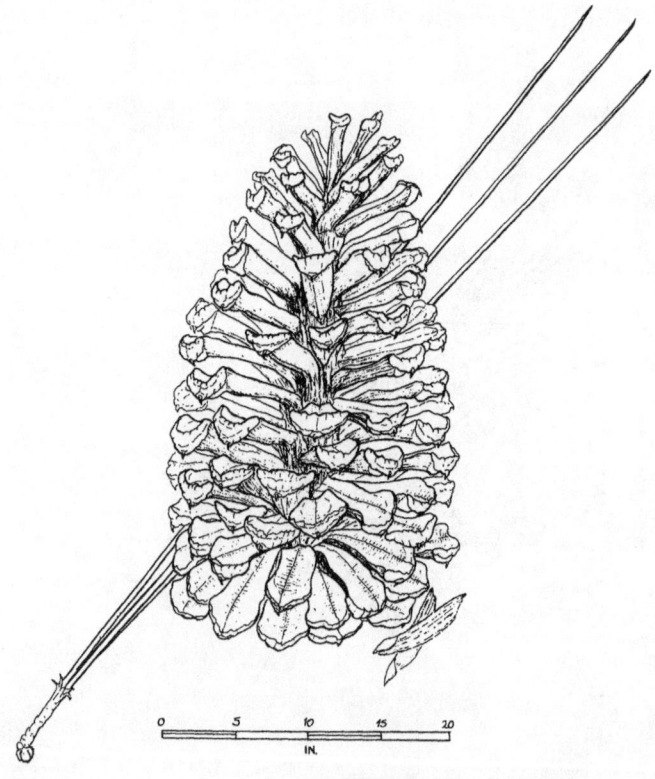

Figure 119. Pinus palustris (Longleaf Pine)

needles. In the wild the plants remain in this stage for some time before the trunk begins to elongate further. Under cultivation, the grass stage may be shortened to only one year. The cones are narrowly conical when closed, becoming broadly conical when they open, and are 6–10 inches long. Each scale-face has a recurved spine on it. The longleaf pine is one of the most fire-resistant species and is also quite resistant to diseases.

———*Pinus strobus*, WHITE PINE, EASTERN WHITE PINE (fig. 120). It is surprising to find the white pine growing in Zone 8 because it is definitely at home much farther north, the southern limit of its natural range being the slopes of the Appalachian Mountains. But it does occur in Zone 8, though it is rarely grown. This tree is more commonly associated with the northeastern United States, calling to mind the poem *Evangeline* by Longfellow:

This is the forest primeval. The whispering pines and the hemlocks . . .
Stand like Druids of old.

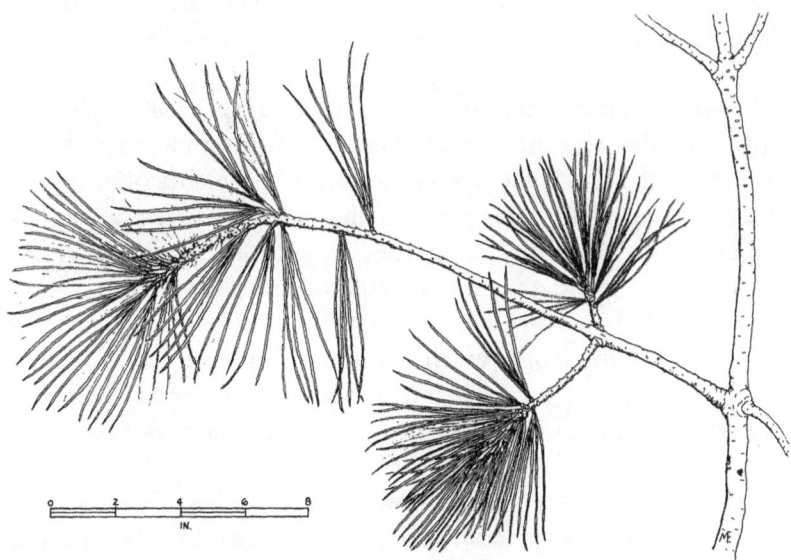

Figure 120. Pinus strobus (Eastern White Pine)

The trees in their native regions may grow to heights of 120 feet, though trees of this stature are a rare occurence today because most of the original stands have long since been cut. The white pine is a beautiful tree, with silvery-green, flexible needles, five to a bundle, 3–5 inches long. The basal sheath of the needles falls off early in leaf development, unlike those of many other pines where the sheath remains throughout the life of the needles. The bark of the younger branches of the tree is quite smooth, and the scaly bark associated with most southern pines occurs only in the older stems. We have not seen cones on the trees growing in the South, but they are commonly 4–8 inches long, narrowly oblong, often somewhat curved. The scales of the cone do not have spines, as is true with many other southern pine species. Growth of this tree in Zone 8 seems to be quite rapid. It is probably best to plant the white pine on the north side of the house or other protecting structure where it will receive a lot of shade. This zone's (mostly) acid soils are fine for the tree since this condition is found in its native habitats. The white pine is a fine tree for lumber. The wood is soft and easy to work and was a favorite of the furniture makers of the early colonial days.

————*Pinus taeda,* LOBLOLLY PINE, OLD FIELD PINE (fig. 121). This is another tree that is sometimes planted for pulp, though not as often as the slash pine. Its needles occur in bunches of three, 6–9 inches long. The cones are 3–4 inches long, narrowly conical-cylindrical when open. Though the lumber from this species is not of the highest grade, the timber is still sold as yellow pine.

————*Pinus virginiana,* VIRGINIA PINE, JERSEY PINE, POVERTY PINE (fig. 122). As the last name indicates, the plants do very well in impoverished soils. Although the Virginia pine is not native to Zone 8, the species is offered by the Florida Forestry Service to be planted in deep sand regions, particularly in north Florida. The plants are much shorter at maturity than some of the other species, 20–50 feet tall. The species' predominant use is for pulpwood, but it can be used for Christmas trees. The needles are in bundles (fas-

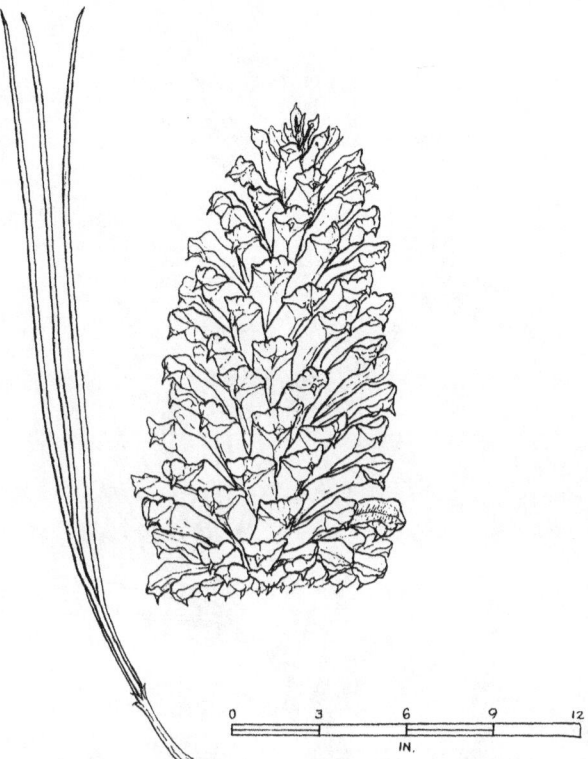

Figure 121. Pinus taeda (Loblolly Pine)

cicles) of two, twisted, up to 3 inches long. The cones are very small, 1½-2½ inches long, and are conical.

PITTOSPORUM. The species in this genus come from various warm temperate and tropical areas of the Old World. They are all evergreen and quite a few can be used as ornamentals. They are most frequently raised for their leaf show rather than for their flowers, though the flower heads do make a fairly nice show. These plants are mostly used as background plants, but some of them, particularly the variegated forms, may be used as display plants by themselves. They take to pruning nicely and can be made into fan-

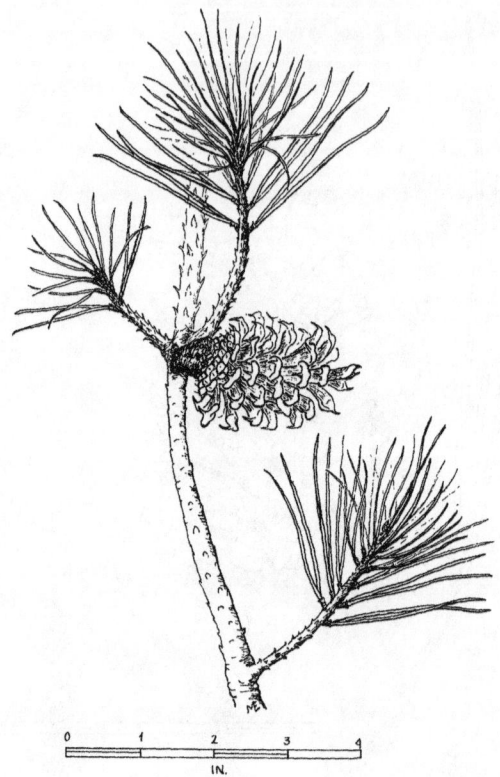

Figure 122. Pinus virginiana (Virginia Pine)

ciful shapes, if desired. The very cold weather of the late 1980s has been devastating to these plants.

————*Pittosporum tobira,* JAPANESE PITTOSPORUM, AUSTRALIAN LAUREL (fig. 123). None of these common names is heard in our area, particularly the last listed, which generally calls to mind another group of plants entirely. We usually refer to them by their generic term, "pittosporum." Varieties of this species have leaves ranging from dark, glossy green through light green to variegated green and white. All are attractive, and most are quite hardy in this zone. They are easy to cultivate and can be grown from cuttings. The plants will grow quite tall (it is said up to 18 feet) but, as mentioned above, are usually pruned regularly to lower heights. This

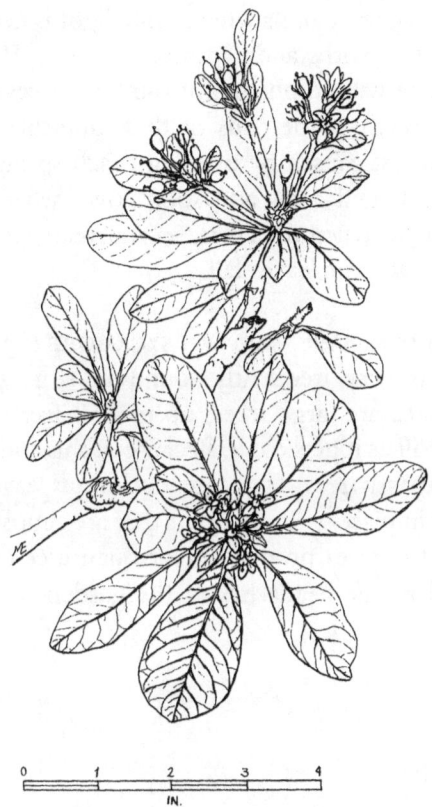

Figure 123. Pittosporum tobira (Japanese Pittosporum)

species is particularly useful to landscape specialists as decoration in the large shopping malls or greenery around restaurants and other commercial establishments.

PLATANUS. Species in this genus, the sycamores, the plane trees, or buttonwoods, are found in most parts of temperate zones, north and south, grown along streets or in large yards. They grow to over 150 feet tall, have wide-spreading branches, very handsome bark, and large, deciduous leaves. Though the custom is not frequently practiced in the United States, the sycamore is the street tree of choice in many European cities not only because it is deciduous (a great benefit in colder regions where more light is needed in winter) but because it can withstand repeated pruning. Walking along a street in France in winter, one might think the trees have some sort of tumorous growth at the ends of their branches. But these enlarged areas are just caused by pruning. Each spring, many shoots arise in these enlarged areas and soon cover what is otherwise a rather unattractive structure. Then, several years later, the tree will be pruned again after its leaves fall.

———*Platanus occidentalis,* EASTERN SYCAMORE (fig. 124). The native sycamore is very frequently found in large yards or along streets. Sycamores are large trees, up to 150 feet tall, with white bark that strips off as it gets older. Because of this and other features, they are rather messy trees. The large, deciduous leaves all fall at one time, and if left in place will form mats that are almost impenetrable. In the spring, the trees produce great quantities of round fruits, which also add to the headaches of the gardener. Having said all

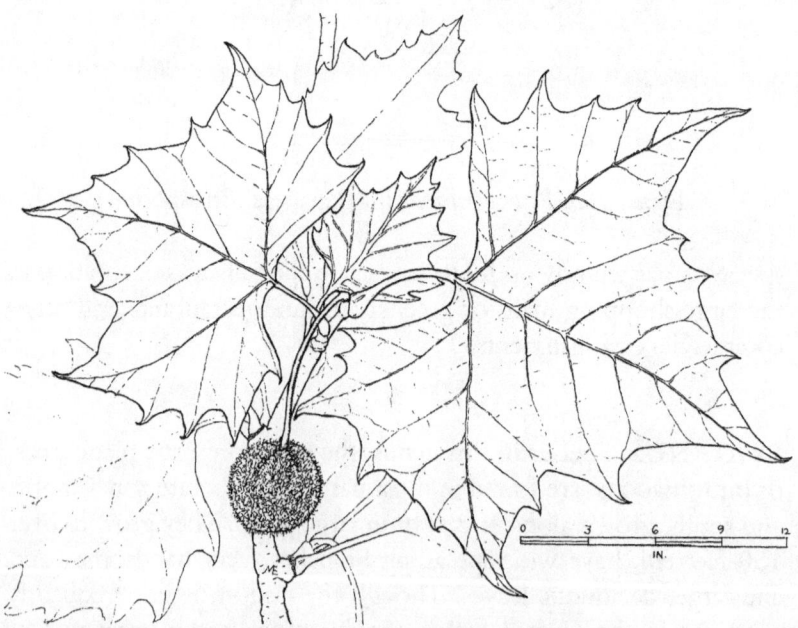

Figure 124. Platanus occidentalis (Sycamore)

these bad things about this tree, it is still one of the most attractive large trees of the Deep South and is worth the trouble it causes.

PODOCARPUS. This genus contains about seventy-five species. These evergreen trees are more often treated in Zone 8 as shrubs, a form maintained by pruning. These plants are conifers (cone-bearing), a group that includes pines and cypress. They are native to the more temperate regions of the southern hemisphere. They are included here for their value as foundation plantings, used either individually or in groupings. *Podocarpus* is sometimes tender in the Deep South but usually it is hardy.

———*Podocarpus macrophyllus,* and the dwarf variety, *P. macrophyllus* var. *Maki,* SOUTHERN YEW, JAPANESE YEW (fig. 125). The dark green, somewhat glossy, 1–3 inch leaves are strap-shaped and grow densely along the stems. Once established, the plants grow without much care, except for the necessary pruning to obtain the size and shape desired. As stated, they can be somewhat tender in Zone 8. They may be accent plants, one standing alone at a corner, or one on each side of an entrance. This plant also makes a good hedge, particularly the dwarf variety.

POPULUS. The poplars, aspens, or cottonwoods, as various members of this genus are commonly called, are found more frequently north and west of Zone 8. In the mountainous west, the aspens are very striking, being one of the most frequent broad-leaved, deciduous plants in otherwise coniferous (pine, spruce, or fir) forests. In the plains, at lower elevations, the cottonwoods are the most evident tree species. All species of *Populus* are dioecious (sexes occur on separate plants). Most of the cottonwoods planted by humans are male plants because the seeds on the female trees give them an unsightly appearance. The light, hairy seeds are produced in very large numbers per tree. When they are mature, they are shed in such enormous quantities that they become a nuisance, forming fuzzy bunches in every nook and cranny, inside and outside the house. We

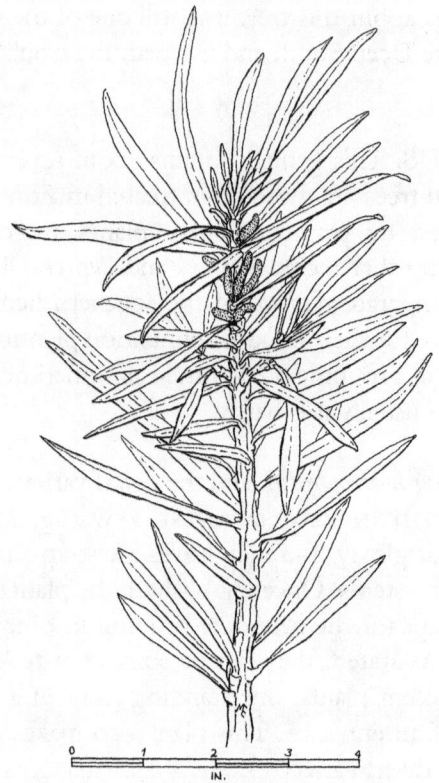

Figure 125. Podocarpus macrophyllus (Japanese Yew)

do not see this phenomenon in the one representative of the genus found in Zone 8 because only the male trees are propagated.

————*Populus nigra* cv. Italica, LOMBARDY POPLAR (fig. 126). This handsome, columnar tree apparently arose as a mutant in the Italian region of Lombardy. It has been in cultivation for a long time, sometimes lining driveways, sometimes planted as a tall windbreak for other crops, and sometimes used as an individual specimen plant for accent. The deciduous leaves are wedge-shaped, dark, and shiny green. The plants grow rapidly to heights of 50–60 feet and thus are particularly useful as plantings around new homes devoid of any landscaping. Unfortunately, the trees die back in ten to twenty years.

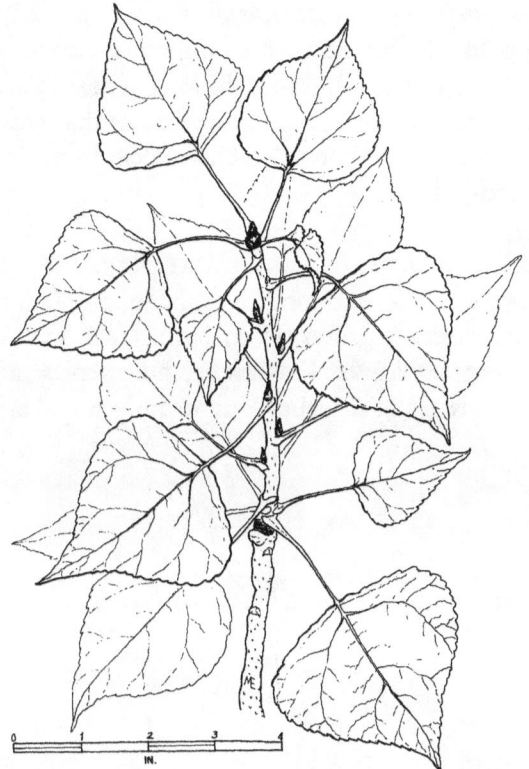

Figure 126. Populus nigra cv. Italica (Lombardy Poplar)

PRUNUS. There are many common names for members of this genus, such as sweet cherry, cherry laurel, chickasaw plum, peach, and sloe (either *Prunus alleghaniensis* or *P. americana*). There are some 400 species in this genus of trees and shrubs worldwide.

Species native to Zone 8.

———*Prunus americana* or *alleghaniensis,* SLOE. This is a small tree, with very early (February) flowering, and is very fragrant. It is infrequently found in cultivation in this area.

———*Prunus angustifolia,* CHICKASAW PLUM (fig. 127). This is a small tree up to 20 feet tall, with numerous, slender, spreading branches from a short trunk. White flowers appear in early spring, March–April, followed by red or yellow fruits that are ½ inch in diameter. They are good for jelly. This species is seldom seen in gardens or yards.

———*Prunus caroliniana,* CHERRY LAUREL (fig. 128). The cherry laurel is one of the most common trees in Zone 8. It is most often used as a shade tree. It grows up to 25 feet tall, with a rounded crown and evergreen leaves. Flowers are on racemes (a stalk with many flowers), are up to 3 inches long or longer, and are followed

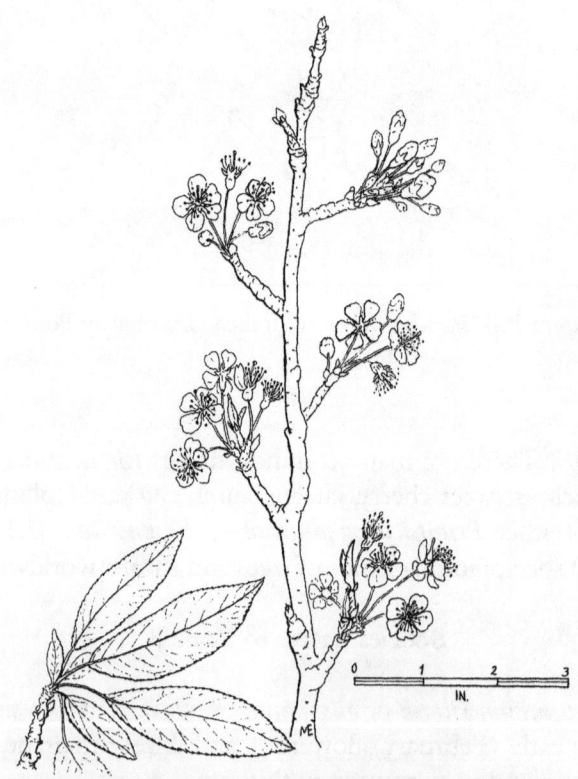

Figure 127. Prunus angustifolia (Chickasaw Plum)

Figure 128. Prunus caroliniana (Cherry Laurel)

by many small, black fruits loved by birds. The plants may be pruned to a hedge or various, individual shapes. This species is a vigorous grower, and thousands of seeds germinate yearly.

Species introduced to Zone 8.

———*Prunus campanulata,* TAIWAN CHERRY (fig. 129). The Taiwan cherry is found in several areas in Zone 8, but in some references is said to occur to the north of this area. The authors first found out about this cherry from a gentleman in Pensacola who has had a couple of trees in his yard for a number of years; he got his start with some plants grown by a friend of his in Fairhope, Ala-

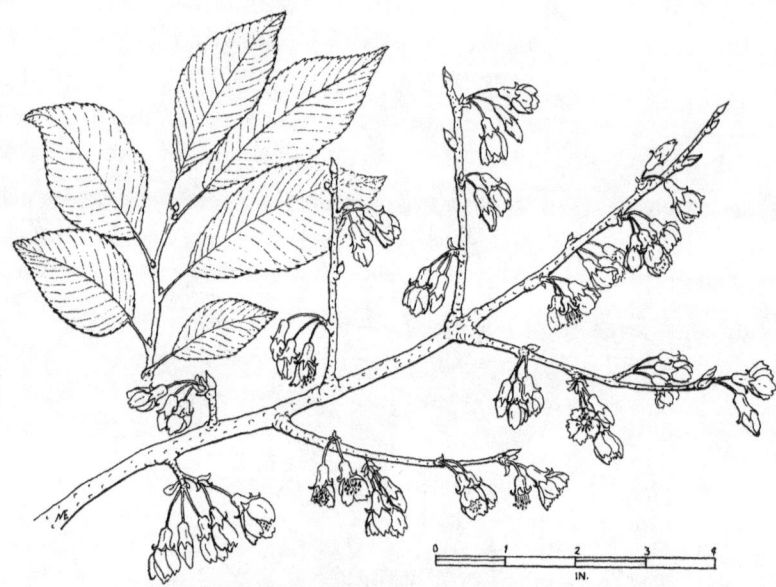

Figure 129. Prunus campanulata (Taiwan Cherry)

bama. In Pensacola, the trees had grown to a height of 25–30 feet, which may be their mature height. The plants have been in place long enough that it appears they can withstand the weather extremes of Zone 8, at least the cold and probably the hot as well. The Taiwan cherry is one of the earliest flowering trees in this area, flowering before the leaves come out from late January into middle or late February. The handsome, dark red flowers hang in numerous clusters from branches all over the tree; they are mostly single but a few are double, giving a pleasing variability to the tree. The younger stems have numerous, prominent lenticels or pores scattered along the upper and lower surfaces. The leaves are prominently serrate, 3–5 inches long, elliptic to slightly ovate, and, as mentioned earlier, they develop after the flowering period. The small fruits fall to the ground at maturity and numbers of plants germinate from their seeds. The offspring are true to color and provide an easy way for you to propagate more plants either for yourself or for others. This is a very fine tree that deserves special placement in your yard. They grow in sun or light shade and thrive under normal planting, watering, and fertilizing activities. They need water at least once a week until they are fairly well established.

————*Prunus glandulosa* cv. Sinensis, FLOWERING ALMOND, MARCH ROSE (fig. 130). This small, delightful shrub, which flowers in early spring, has probably been in Zone 8 gardens for a very long time. It does not seem to be very popular, however, as it is rarely seen. The flowers are light pink and are distributed along the main, upright stem for most of its length, seeming to come directly from the stem without any side stems below the flowers. The flowers are double; thus it isn't easy, at first, to tell that they are really like the flowers of other members of the same genus. The plants grow in

Figure 130. Prunus glandulosa cv. Sinensis (Flowering Almond)

clusters arising from common underground stems (rhizomes). They are 2–3 feet tall. We have never seen any fruit on the plants; it may be sterile. Propagation is possible by underground rhizomes.

———*Prunus persica,* FLOWERING PEACH. This is one of our most handsome early-flowering, small, deciduous trees. Growing to about 25 feet tall, these trees have typical peach leaves, 3–6 inches long, lance-shaped, with serrulate margins. The flowers are about 1 inch across, and they appear before the leaves. This species may bear double or single flowers that are colored red, pink, or white. Flowering peaches have rather small, hard, fruits about the size of the last joint of the thumb. The double-flowered types are labeled according to a special system of nomenclature; their names usually have a latinized ending, "*-plena,*" meaning doubled, in combinations with a Latin word indicating their color. Thus *alboplena* indicates the white double form. These trees grow in light shade or open sun in soils with plenty of organic matter (such as compost, peat moss, or cow manure) and with fertilizers of 8–8–8 composition, preferably slow-release types. Be sure your fertilizer has a good balance of the micro elements (boron, zinc, iron, etc.) as well as the three major components.

———*Prunus serrulata* 'Kwanzan,' JAPANESE CHERRY, ORIENTAL CHERRY. The Kwanzan cherry grows from Zone 8 north to Zone 6 (refer to zone map in frontispiece). This is a tree, up to 60 feet tall, flowering just before or with the young leaves in mid-spring, from April to May. The flowers are light pink to white and occur in great profusion, making a handsome show. The plants seem to thrive in our zone and should be placed as an accent or in a border where they may be displayed well.

❦

PUNICA. The pomegranate is undoubtedly one of the oldest plants cultivated for its fruits. It is mentioned in the Bible and other ancient literature from the Mediterranean area and eastward into southwestern Asia. There is but one species of the genus *Punica,* with many variants. By far the most important variant is the one raised for its fruit.

These plants are interesting not only because of their flowering and fruiting qualities but also as a botanical curiosity. Botanists cannot quite figure out the relationships of the genus and family. The fruit itself, which is sometimes called a berry, is also quite a botanical oddity. At maturity, the fruit has a leathery rind, brownish to orange-red, with numerous, sweet-acid, red, fleshy seeds scattered seemingly indiscriminately all through the fruit inside of papery, yellow divisions. However, if one has the patience to follow the development of the fruit from the flowering stage to full maturity, one discovers that the ovary is evenly divided into four compartments, two above and two below. But as the fruit development continues, upward and sideward movements of the compartments cause them to be distorted, and lose any semblance of order.

———*Punica granatum,* POMEGRANATE (fig. 131). This is a shrubby plant or small tree. It is grown worldwide in warmer regions primarily for the fruit, but some of the selections, especially the shrubby ones, are grown for their handsome orange-red flowers, which occur in May–June in Zone 8. The flowers appear first and then the glossy, green, elliptic leaves develop. Fruits do occur on the plants raised mostly for their flowers, such as *P. granatum* cv. Nana. These fruits have the same shape though usually they're a little smaller than the plants raised specifically for fruit production. These very nice shrubs will do well in most Zone 8 soils, either in light shade or in the sun. We would recommend that they be used in locations where their leggy lower stems are not seen.

PYRACANTHA. This is a small genus of about six species in the rose family, a family that contains many beautiful and useful plants. The pyracanthas are shrubs or small trees with thorny branches and mostly evergreen leaves. There are four main species that have been cultivated for a long time in their native regions of Europe and Asia, and these species have several developed cultivated varieties (cultivars). The plants may be raised as free-standing shrubs or trained up a lattice or wall (espaliered). It would be difficult to tell which variety of which species is used the most in Zone 8, but the following are likely to be found in this area. One problem that should be

172 ❧ *Pyracantha*

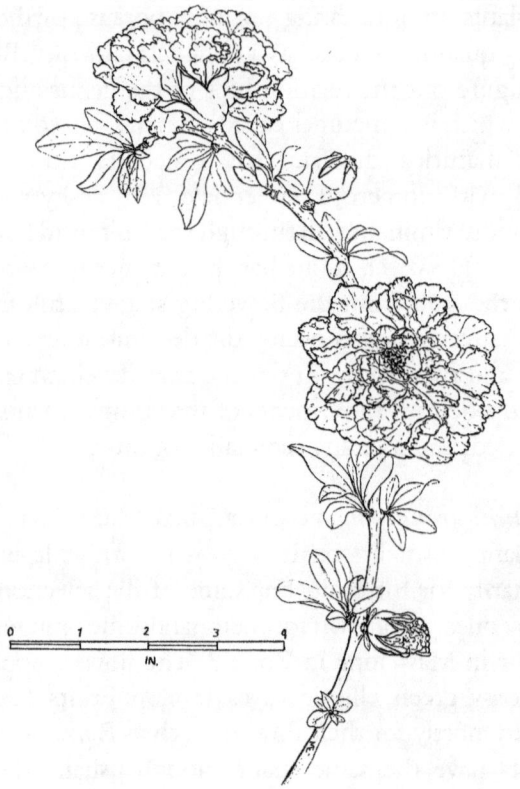

Figure 131. Punica granatum (Pomegranate)

recognized is the "fireblight," a bacterial disease that causes leaves and fruit to look scorched and brown while hanging onto the plant. This can be controlled with sprays. Consult the Farmer's Co-op Store or garden supply stores for products to use.

———*Pyracantha coccinea*, FIRETHORN (fig. 132). This common name is used for all the species. The species name *coccinea* means red or crimson in Latin. It refers, of course, to the bright fruits of the plant, and these fruits are its major attraction. In the late summer, lasting well into the fall, the plants provide a good show from the large numbers of fruits (which in structure are quite similar to the crabapple), each fruit about ¼–⅜ inch across. These can be propa-

Figure 132. Pyracantha coccinea (Firethorn)

gated from seed, but no prediction can be made about whether the seeds will produce the same cultivar that produced them. Cuttings or graftings are the best bet for propagation; they root quite easily from "ripe" wood, which means mature but not too old.

———*Pyracantha crenulata*. This species (also called firethorn) is a bit larger than the preceding and the fruits are perhaps a bit more orange red. Other characters that separate the species are technical.

❀

PYRUS. The pears are included here because they really do make a good spring show with the many white flowers that appear before the leaves. Most of the choicest edible pears do not grow well as far south as Zone 8. This is unfortunate, because no fruit is more delicious when fully ripe. There is an interesting practice of cultivation of pears (mostly in Switzerland) where the young pear is inserted in a clear glass bottle which is attached firmly to the stem. When the pear is mature the stem is severed and the bottle is filled with brandy. After some time, this makes a really elegant cordial.

———*Pyrus calleryana,* BRADFORD PEAR. There are several other cultivars of this species, which is raised solely for its decorative, white, showy flowers in early spring. The flowers have a rather unpleasant odor so the tree should not be raised too close to the house. The leaves, similar in size and shape to other pears, turn a nice red in the fall. Some of the variants, including Bradford, are reported to be susceptible to fireblight, so care is necessary in selecting the plants you wish to raise.

———*Pyrus pyrifolia,* SAND PEAR (fig. 133). This tree has been cultivated for a very long time in these parts, though its origin is somewhere in the Orient. There are some selections of this species that have less of the grittiness that gives the species its common name. Fruits are used primarily for making pear preserves, pear butter or pear pies.

Dr. Frederick Meyer says (personal communication): "Le Conte's Pear, *Pyrus* × *lecontei (P. pyrifolia* × *P. communis)* is the one most commonly seen in the Deep South. The true *pyrifolia* may not be all that common, if, in fact, it can be truly identified!"

❦

QUERCUS. The oaks and the pines characterize the Deep South more than any other vegetation. Botanists and foresters all use the combined term "oak-pine forest" to designate the region. We know how important the oaks are in our landscape, both aesthetically and commercially. The spreading giant live oak has few equals in the plant kingdom for sheer majesty. These same oaks were one of the earliest commercial resources in the southern economy, providing beams, planks, ribs, and other timber for sailing ships. Indeed, the

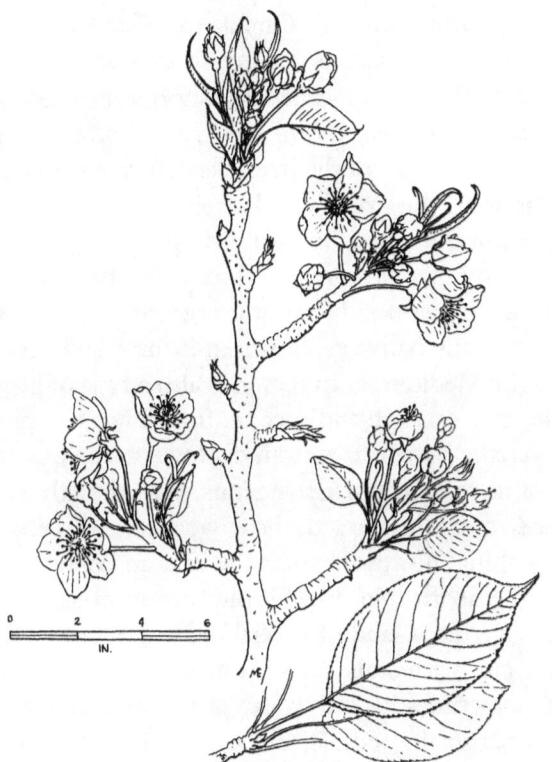

Figure 133. Pyrus pyrifolia (Sand Pear)

U.S. Navy early knew that it had to protect this magnificent tree from uncontrolled exploitation by establishing a live oak reserve on Santa Rosa Peninsula, near Pensacola, Florida. This area is now a part of the Gulf Islands National Seashore and is called simply Naval Live Oaks. The furniture industry would be much worse off if there were no oaks. Today, sadly, there is no restriction against exploitative developers of shopping malls and high-rise condominiums, who see only the raw dollar value of each square foot from which they can make a profit. Just recently, there was an effort in Pensacola to prevent total destruction of the oaks and magnolias from wanton development, with some, but insufficient success.

The world has many oaks, 450 or more species, all in the northern hemisphere, from southern Canada to Costa Rica and from Europe across Asia to Japan, occurring from seashore to mountain side. Many are stately trees, others are shorter, gnarled and twisted (as are southern scrub oaks). Still others are shrubby or small trees, such as the coastal forms of the live oaks that hug sand dunes. Their uses are manifold; they yield oak barrels, corks for bottles, tannin for curing leather, lumber for furniture, panelling, and flooring— on and on. Many species serve even today for firewood.

The oaks are either deciduous or evergreen, but most species are deciduous; the only native evergreen in Zone 8 is the live oak. The cork oak of the Mediterranean region is also a type of live oak as are many of the tropical highland species. It is difficult to pinpoint just one characteristic by which we can distinguish the oak at a glance, but many of us can recognize the genus, even though we would be hard-pressed to say just exactly how we knew a tree was an oak. There is one thing, however, which the oaks and only the oaks have, and that is the acorn. No other group produces such a structure. The acorn is a fruit, a kind of specialized nut, attached to the stem by way of a cup with scaly sides. Sometimes the cup encloses the whole nut, sometimes it's only a cap at the stem end. Some of the acorns are (marginally) edible, but they require processing to rid them of the high concentrations of tannin. The following list of oak species in Zone 8 is certainly not complete and only some of those likely to be used in yards have been listed. There is one species introduced from Asia that has found its way into our gardens, *Quercus myrsinifolia,* an evergreen that deserves more consideration than it has received.

———*Quercus falcata,* SOUTHERN RED OAK, SPANISH OAK (fig. 134). This handsome oak, 70–80 feet tall, its trunk growing to be 2–3 feet in diameter, its crown rounded, is one of Zone 8's nicer species. Its leaves are deciduous, 5–9 inches long and 4–5 inches wide. The leaves are variably lobed, from shallowly three-lobed to deeply five- to seven-lobed, all with a characteristic sharp bristle at the tip, including a longer one on the terminal lobe. Although parasitic mistletoe is a common problem for the water oaks, it does not seem to be as frequent in southern red oaks.

Figure 134. Quercus falcata (Southern Red Oak)

————*Quercus laurifolia,* LAUREL OAK, WATER OAK (fig. 135). The laurel oak is very frequently found in the Deep South. It and *Q. nigra,* which is more appropriately called the water oak, make up the largest number of oaks to be seen in cities and towns. The laurel oak grows quite rapidly, getting to be a large tree up to 70 feet tall. Unfortunately, its life span is quite short, not nearly as long as the live oak, but it still serves quite nicely for an easy-to-maintain shade tree. The leaves usually hang on until late winter or early spring before shedding.

————*Quercus marilandica,* BLACKJACK OAK (fig. 136). This oak is sometimes seen in yards and gardens though it probably is not often

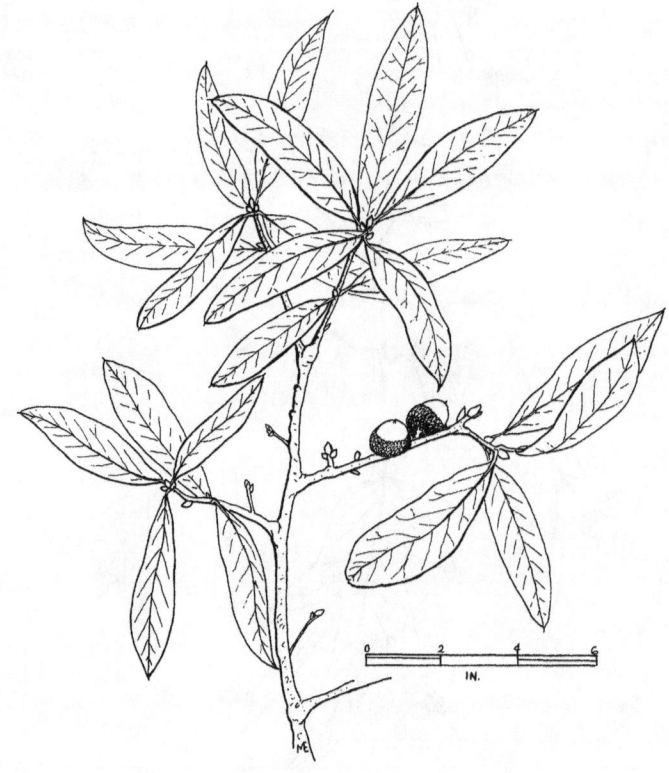

Figure 135. Quercus laurifolia (Laurel Oak)

purposely planted. Its normal habitat is rather sandy areas. The blackjack oak doesn't grow as tall as several other species, often not more than 30–40 feet tall. The bark is sort of blackish and very rough. It has a straight trunk but twisty, spreading branches. Its leaves are not exactly lobed but have slight indentations on each side of the main axis at the top of the leaf. The acorns take two years to mature and the cup covers from one-third to one-half of the whole length. One advantage of this oak over, say, the water oak, is that all of its leaves drop at about the same time in the fall rather than continuing to fall over a long period.

————*Quercus myrsinifolia*, CHINESE EVERGREEN OAK (fig. 137). This species, though known to botanists for a long time, does not

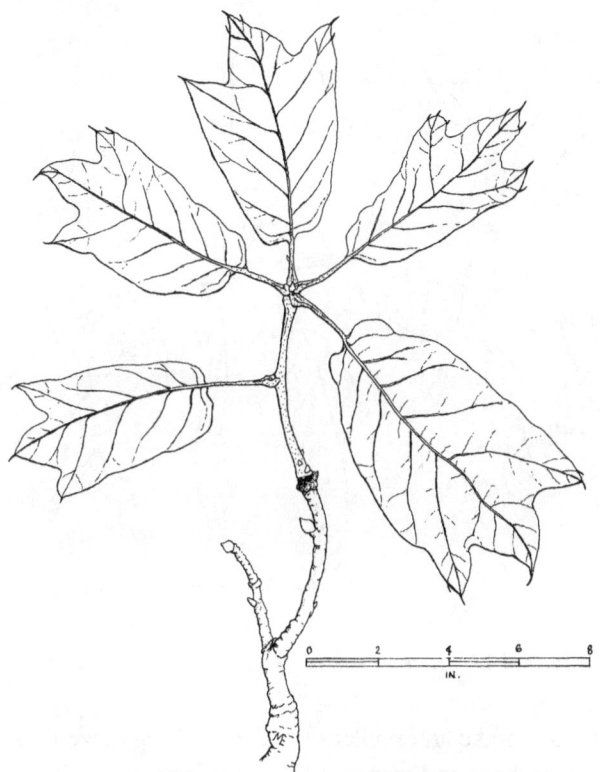

Figure 136. Quercus marilandica (Blackjack Oak)

seem to have been widely recognized in the horticultural trade, at least not for our area. Most references indicate that it grows to the north of Zone 8 but not this far south. That it does grow here is evident—we do not include plants we haven't seen. As stated in the generic discussion above, it is evergreen, and the only oak from Asia we have discussed. *Q. myrsinifolia* is said to reach about 60 feet, but we do not know its shape at maturity. Plants to 20 feet tall are narrowly conical in outline. The leaves are shiny, up to 5 inches long, with an elongated, narrowing tip. The leaf margins are toothed. We have not seen the acorns, but they are said to have a cup enclosing one-third to one-half the nut. There seems no doubt of this species' reliable winter hardiness, and it seems to thrive in the hottest of summer weather. Because of its shape and its evergreen leaves, the

180 ❧ *Quercus*

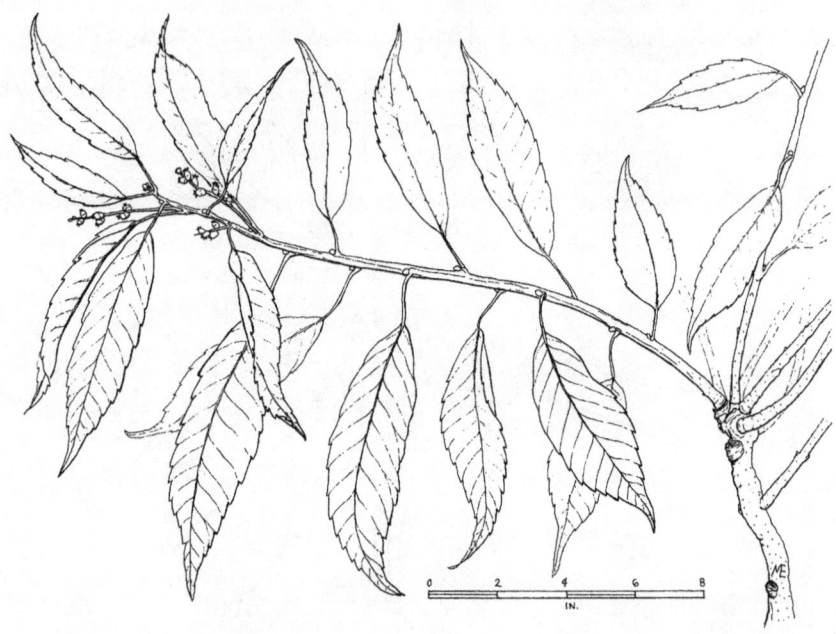

Figure 137. Quercus myrsinifolia (Chinese Evergreen Oak)

plant would make an excellent border for long driveways or may be used as a background tree in a large garden.

————*Quercus nigra,* WATER OAK (fig. 138). The water oak is distinguished from the laurel oak by having leaves that are broad at the upper end and have smooth tips, whereas the laurel oak leaf is more nearly elliptic and has a small point on the tip. Otherwise, the characteristics of the two species are similar. Both grow rapidly and are relatively short lived. Both are susceptible to infections of mistletoe.

————*Quercus virginiana,* LIVE OAK (fig. 139), and *Q. virginiana* var. *maritima,* SAND LIVE OAK. The stately live oak, with its rounded crown, wide-spreading limbs, and evergreen leaves, is almost as much a trademark of the South as is the southern magnolia. This species, which finds its greatest development not too far from salt water, grows best in the coastal areas, though it will do quite well inland. The live oak, including the variety that grows in deep sand,

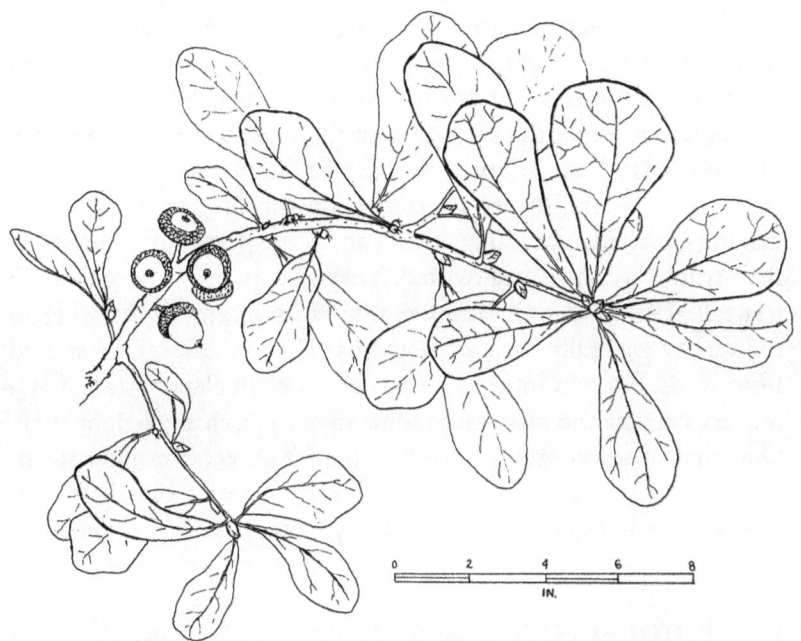

Figure 138. Quercus nigra (Water Oak)

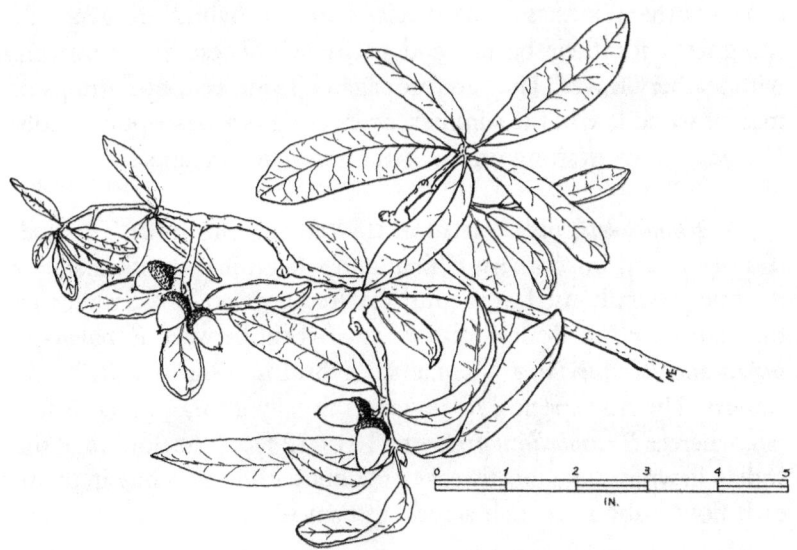

Figure 139. Quercus virginiana (Live Oak)

most frequently just behind the foredunes of coastal areas, the sand live oak, is the only evergreen species native to Zone 8. The leaves are leathery, linear to narrowly ovate, 2–4 inches long, are shiny green above, lighter green and sometimes downy beneath. Characteristically, the leaves roll downward and inward, "revolute" in botanical terms, along the sides. About the main distinction between this species and its variety is the size of the mature plants: in the coastal variety, the plants are frequently shrubby, with stiff twigs making a matted appearance above, and if the plants grow into trees their trunks are small and twisted. They seldom grow more than 15 feet tall. The species *Q. virginiana* bears its acorns singly, whereas the variety generally has paired acorns. The live oak can be started from seeds, but it is much better to buy a plant already started at a nursery because the plant takes some time to reach any height at all. One plants the live oak for posterity; don't expect overnight splendor from these slow-growing trees. But do plant one, giving it plenty of room to develop for the enjoyment of the next generation.

❦

RHAPHIOLEPIS. The rose family, Rosaceae, to which this small genus belongs, has many genera and species that are used in ornamental horticulture. *Rhaphiolepis* is an entirely Asiatic genus, with some fourteen species. Two species and one hybrid are generally recognized for their beauty and usefulness. These are evergreen, with leathery leaves. They are propagated from seed, by cuttings of mature wood late in the summer, by layering (see description under *Deutzia*), or by grafting on stock of *Crataegus* (crabapple).

———*Rhaphiolepis indica,* INDIAN HAWTHORN (fig. 140). This species originated in southern China, and while the common name is the one generally used in the nursery trade, we are not certain that the plants used in Zone 8 may be the hybrid between *Rhaphiolepis indica* and *R. umbellata.* The named hybrid is designated *R.* × *delacourii.* The first species, *R. indica,* is usually no more than 5 feet tall, whereas *R. umbellata* grows to 10 or 12 feet. The flowers of the Indian hawthorn are mostly rosy pink but with some white in them, each flower about ¾ inch across, in paniculate clusters. Flowering

Figure 140. Rhaphiolepis indica (Indian Hawthorn)

begins in May. The shrubs may be used as background plantings in borders or singly as accent plants. They seem to be hardy in Zone 8.

RHAPIDOPHYLLUM. This is a genus in the Palmaceae, the palm family, with only one species. This is a fan palm, in which the blade spreads out like a fan from the petiole or leaf stalk.

————*Rhapidophyllum hystrix,* NEEDLE PALM, PORCUPINE PALM (fig. 141 a and b). The needle palm has long been known in cultivation and has the distinction of being able to withstand colder weather than almost any other palm. Its native range, from South Carolina around the coastal plain to Mississippi, is no greater than

Figures 141 a and b. Rhapidophyllum hystrix (Needle Palm). Plants to 6 ft. tall. 141a–Shiny foliage. 141b–Base of plant with sharp spines.

is that of the cabbage palm (*Sabal palmetto*) which is also a native species. Within this range the plants have a rather sporadic distribution, found only in very shaded, heavily wooded locations. If found at all, it is usually near the bottom of steep-sided ravines. The needle palm grows only to about 5 feet tall, and grows in clumps or clusters linked together by horizontal stems just at or under the soil. Its foliage is glossy dark green, and very sharp needles 5 or 6 inches long surround the base of the leaves. One uses caution when digging around this plant! Needle palms grow easily in shady areas with little more care than frequent watering. A dusting with an insecticide is advised when the plants are young, directly on the growing points to keep down a type of caterpillar that will kill the growing point.

❦

RHODODENDRON. The azaleas are included here, in keeping with the latest scientific thinking. Some of the world's most beautiful shrubs and small trees belong to this very large genus—perhaps 800 species, occurring in all continents except Africa and South America. Most think of these plants as shrubs, so it comes as something of a surprise to hear that the genus includes small tree species. Some beautiful *Rhododendrons* are native to the Southeast, from Texas eastward through Tennessee, and from southern Pennsylvania to Virginia and southward to about the middle of Florida. Although there are a few evergreen species, particularly in the mountainous parts of the Southeast, the most familiar species are those that drop their leaves for at least a part of the year. Some of these latter species are quite fragrant, which leads to common names such as "honeysuckle." There are three basic colors in the southeastern, native, deciduous types: yellow, pink, and white. There are variations in intensity of these colors and differing shades of each, and some are actually bicolored. The Deep South (Zone 8) is generally too warm for the evergreen species that make such spectacular shows farther north. *Once again, we emphasize that gardeners should not take the native species from the wild. Many nurseries are well equipped to germinate and grow seeds taken from wild plants, and this is the most acceptable procedure.*

The cultivated azaleas that make such a beautiful appearance in our gardens starting in March, finishing some time in April, are mostly evergreen and are more likely hybrid species than "pure" species. There are two hybrid groups that account for most of the plants in this area: the *indica* and the *kurume* hybrids. The indicas have large flowers that generally appear before the spring leaf flush; the most commonly found color is magenta, although there are white and yellow ones as well. The kurume types are mostly dwarf and small-flowered. We can apply many other names and not too many of our local nurseries will know all possible variations. No one should pick on the local nurseries for this fault, however, because few people are familiar with this information, the authors included! Those who would like to become more familiar with these names may write to the American Horticultural Society in Mount Vernon, Virginia. This organization can provide lists of their own and other publications that specialize in the cultivation of rhododendrons.

The rhododendrons grow best in soils that are well-drained, with considerable amounts of organic matter included (such as leaf mold, rotted cow manure or peat moss). For continuing spectacular shows, one should be sure to use a fertilizer specially formulated for camellias and azaleas (or use a good brand of fertilizer with the formula of 8–8–8 plus a good balance of the microelements required). Generally it is recommended that fertilizer be added just after the blooming period in the spring, and again in October–November. If you receive gifts of potted azaleas in the winter, it is best to keep them in their pots until spring, setting them out after the last frost. Watering is critical to young azalea plants, and they should not be allowed to dry out. A good mulch on them is very useful to keep the soil moist and weed-free. After several years of care, the plants can begin to fend for themselves, except during the droughty periods that frequently occur in Zone 8.

The native species.

————*Rhododendron alabamense*, ALABAMA AZALEA (fig. 142). This native is frequently offered by nurseries, though the natural range is

Figure 142. Rhododendron alabamense (Alabama Azalea)

restricted to Alabama. It has white flowers that appear with the leaves, perhaps a bit later in the season than the other natives listed here. The flowers, with some fragrance, are tubular with flairing lobes that are shorter than the tube. The bush is rather low-growing and requires about the same conditions as most azaleas: light shade, sandy soil with a good supply of humus, regular, deep watering, and fertilization at least twice yearly, the first applied after the blooming period, the second in the fall.

————*Rhododendron austrinum,* FLORIDA FLAME AZALEA (fig. 143). The natural range of Florida flame is limited to northwest Florida only. The bright orange to yellow flowers come early in the spring before the leaves, along with the flowers of most of our other native

188 ❧ *Rhododendron*

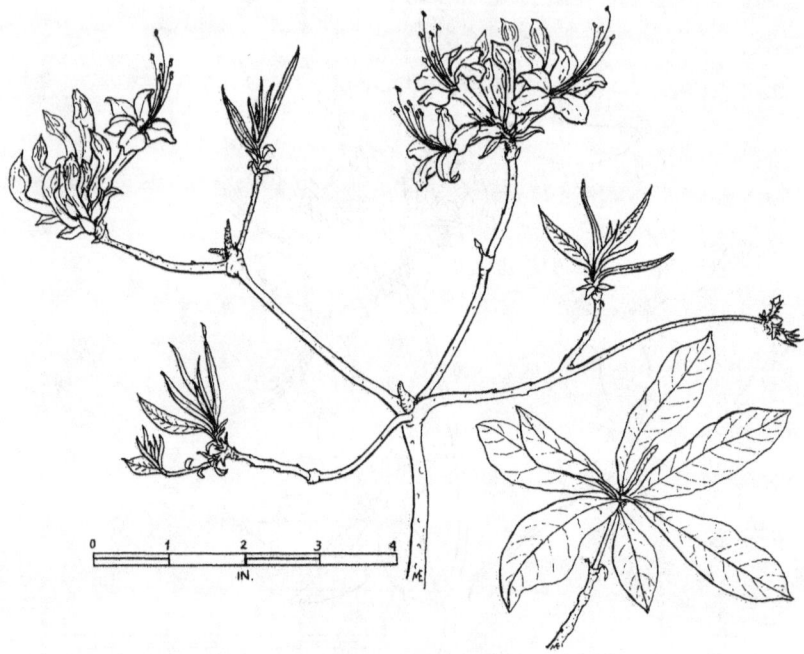

Figure 143. Rhododendron austrinum (Florida Flame Azalea)

azaleas; they are sweet-scented and appear in clusters at the tips of the stems. This is one of Zone 8's showiest plants, and well worth planting. Care is similar to the above.

———*Rhododendron canescens,* BUSH HONEYSUCKLE, PINXTER FLOWER. This is a very popular, pink-flowered species. The plants are native in our area. Some of the plants have flowers that are almost entirely white, with only a little pink in them. They are quite sweet-scented, accounting for the name "honeysuckle." This is a deciduous type, flowering before the leaves, and its cultivation is similar to the other species.

———*Rhododendron chapmanii.* CHAPMAN'S AZALEA. We have not seen this Florida native ourselves, but it is said to be worthy of cultivation. It is found in flatwoods and on the edges of swamps in drainage regions of the Apalachicola River.

———*Rhododendron prunifolium,* PLUM-LEAVED AZALEA. This species is native to only a small area of middle western Georgia, but it has been brought into cultivation, first at Callaway Gardens and then by commercial nurseries. The flowers appear much later than those of most azaleas, sometime in July–August, long after the leaves are formed. The bright, orange-red flowers are a welcome sight in midsummer. The plants may grow to 10 feet tall.

❀

ROBINIA. This is a genus of 20 species and many varieties in the Leguminosae or bean family. Species range from large shrubs to small or medium-sized trees. The flowers are typically like all bean flowers, with two wing petals, a broad standard, and a keel that usually encloses the stamens and pistil, and are either pink or white. The leaves are compound with several to many leaflets, and they are deciduous. Cultivation is from seeds, cuttings, sucker shoots, or runners, and the plants do not require any special type of care. They grow in a variety of habitats, from shady to bright sun. Most often plants have short spines along the branches, so it is not a good idea to have the plants too close to your house where they can "reach out and grab you."

———*Robinia pseudoacacia,* BLACK LOCUST (fig. 144). Our only representative of *Robinia* in Zone 8 is this small to medium-sized tree—farther north the tree grows much taller. These trees range over much of the eastern part of the United States, occurring both in cultivation and as natives. The trunks of these trees are much used for fence posts because they are very hard and long lasting. The wood may be turned on a lathe but is seldom used either for furniture or other construction. The leaves provide only light shade since the tree does not have a heavy canopy. The plants can be grown in problem areas that are difficult to water or where the soil is very dry much of the time. Several horticultural forms of the species have been produced, either by selection within the species or by crossing with other species. The flowers are predominantly white, though some forms may be pink, in pendant inflorescences. Bees are attracted to the scented flowers and produce a perfumed honey.

190 ❦ *Rosa*

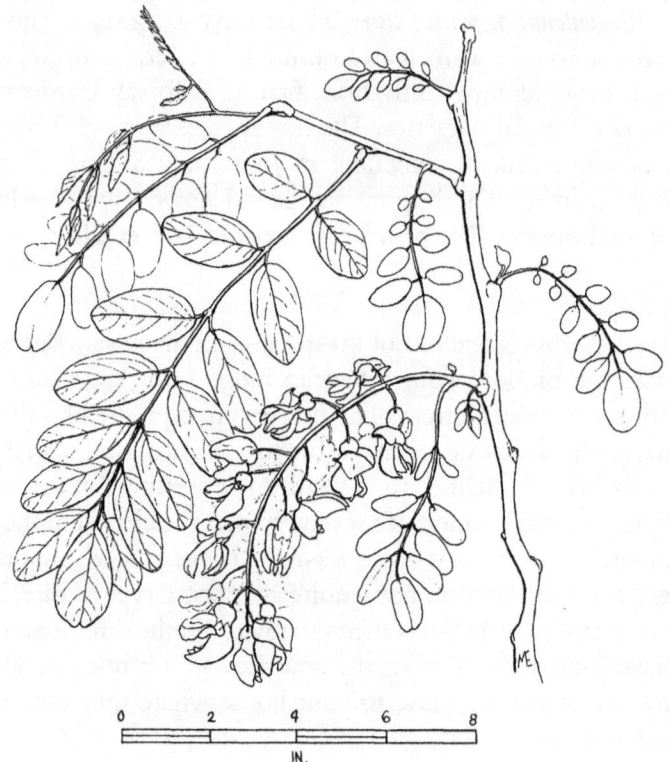

Figure 144. Robinia pseudoacacia (Black Locust)

❦

ROSA. There is little doubt that the roses are the queen of all flowering shrubs, both now and in the distant past, as far back as recorded history. There are many roses, native and cultivated, more than 100 species, all native in the northern hemisphere, and cultivated varieties numbering in the tens of thousands. Now they may be found in the temperate zone of both hemispheres. Wherever temperate zone people have migrated to the tropical areas much effort has been expended attempting to coax roses to grow. But mostly the poor things act as though they want to escape back to where they "ought to be." Amongst the 100 species there is a rich bank of beautiful genetic material that man has, is now, and will use long into the future to produce new and elegant recombinations from old genetic combinations. In spite of the many recombinations, rosarians who have spent much time attempting to know and classify them have been able in most cases to trace them back to the original crosses.

The widespread fascination with roses has stimulated the development of a vast array of specialists and untold numbers of amateurs who share in their love of the plants. The disciplines involved in the specialities cover the same types of scientists and business people as for any other crop of importance: botanical classifiers, physiologists, geneticists, ecologists; ornamental horticulturists, plant breeders, plant pathologists, entomologists, and soil scientists; growers, both wholesale and retail, and nurserymen of the same categories, and, no doubt, others.

For best growth, roses should be raised under open sun, never where tree roots can compete with them. They require regular applications of fertilizers; soils rich in humus from such sources as peat moss, composts, and/or cow manure; regular watering; and at least weekly spraying for diseases and (as needed) insecticides. Soils should be just slightly on the acid side, say a pH of 6.5, though this may vary from locality to locality. Most roses are grown on grafted roots in most vigorous conditions, and the plants should be placed so that the graft is just at or slightly above the soil line. Mulching of roses is important to keep down weeds and to keep the soil cool and moist. In Zone 8, care of roses is a constant—you cannot plant them and leave them and then expect good results. Our high humidity, long growing period, infertile soils (particularly in the near-coastal area), and hosts of fungi and insects put extra demand on the rosarian. As one good rosarian said to us "either be dedicated, or don't grow 'em."

It is important to select good types of roses to get the most rewards. Fortunately, some of the large discount stores now carry good rose varieties, but unfortunately, the clerical staff (usually) has little or no training horticulturally, so the buyer cannot get much advice from them when problems arise. It is more expensive to begin with, but more economical in the long run to buy your roses from a nursery specializing in roses. The staff can tell you what you need to do, and what procedures to follow if you run into problems. There are many reputable rose growers and nurseries, but you must search them out for your own area. Two outstanding areas for roses in the south that we know about are found in southwest Georgia and in northeast Texas, centering on the cities of Thomasville in Georgia and Tyler in Texas.

We should say something about the types of roses, but are loath to do so because the subject is very complex, and there are many experts willing to criticize any condensed statement short enough to fit into this little book. We can recommend very highly the commentary on roses given in *Hortus III* (see "Selected Reading" list), as well as the condensed discussion given by the authors of *Garden Guide to the Lower South,* also cited in "Selected Reading" list. There are many other sources, but anyone who wishes to get the most accurate and complete information would do well to write to the American Rose Society, P.O. Box 30,000, Shreveport, LA 71130.

Not only do we raise roses for their beautiful blooms, but also for their delightful fragrance, and in some cases an excellent jelly (from the fruits, called "hips"). The Damask rose (*Rosa damascena*) is the major source of attar of roses, distilled from the flowers in Bulgaria where most of it is produced. From this attar many expensive perfumes are produced. But along with the beauty and other uses there comes one drawback—the stems are thorny, mostly very sharp. Some are more thorny than others, and no amount of breeding seems to be successful in ridding the plants of this characteristic. But suppose we were successful in a program to breed thornless roses—what would we do without those wonderful sayings about roses and thorns?

SABAL. This is a genus of the palm family and the common name for all of the 20 or so species known as "palmetto." The various species range from North Carolina around the coastal plain to Texas, to Bermuda, Cuba, the north coast of South America, and on around the Caribbean Sea to Mexico. The plants are either erect and columnar, with trunks 60–80 feet tall, or else have trunks that recline, some even bearing subterranean stems. The columnar types have no side branches but grow from a single point at the tip of the stem, whereas the reclining or subterranean types may branch frequently to form a thicket. Cultivation of the plants is quite simple: just dig a hole large enough to accept the roots, cover them with the original soil, support the plants with substantial guy-wires and stakes, and water frequently until the plants indicate they are alive and growing. We seldom, if ever, raise the plants from seed but instead take the fully grown palms from the wild. Quite a business has grown up doing this, and one may frequently see a truck loaded with the palms driving down the highway to markets. Unfortunately, the business is thriving, the demand for the plants very great, and the result is severely denuding the native population.

————*Sabal palmetto,* CABBAGE PALM (fig. 145). This stately tree grows to 60 feet tall and is found mostly in Florida, though it may grow up to North Carolina right on the coast. This species seems to be able to grow in a variety of habitats, but mostly it prefers to grow where the water table is not too far below the surface. The leaves of the cabbage palm may be up to 6 feet long, and are fan-shaped. The trunk of the more mature plants is bare, but the leaf bases of the older leaves hang on after the upper petiole (leaf stalk) and blades have fallen off. Younger plants may have the leaf bases all the way to ground level. The roots of this palm, like those of most other members of this family, consist of pencil-thin, numerous projections around the base of the trunk. It is surprising to see just how few roots support the tree and keep it erect even in hurricane-force winds. This species is at best a subtropical one and, therefore, quite sensitive to cold. The 1986–1990 winters in the Deep South were so severe—particularly from Panama City, Florida, westward—that many of the cabbage palms planted in this area died. This region is, in fact, a bit beyond the range of the plants as natives. Right along the coast, temperatures may be warm enough to prevent loss from cold, but certainly not inland.

Sabal palmetto is similar in general appearance to *Washingtonia filifera,* which takes the place of the former in regions to the west of the cabbage palm's normal distribution. Washington is more cold resistant and thrives in most of Zone 8. The easiest way to differ-

194 ❦ *Sabal*

Figure 145. Sabal palmetto (Cabbage Palm). Plant 30 ft. tall.

entiate between these two palms is by examining the fronds (leaves). The leaf stalk of *Sabal* is smooth, with no spines or teeth along the edges, whereas the leaf stalk of Washington has prominent spines on each side. Another characteristic distinguishing the fronds may be seen in the fans (leaf blades). Fans of the cabbage palm generally curve downward from the base toward the tip, and the two sides of the fan form an open V-shape. In contrast, Washington palm fans project outward and upward and are more nearly flattened. In both species, the older, dead or dying fronds begin to droop downward, eventually drying and forming a skirt of dead leaves. In Washington palms the skirt hangs on for a long time, sheathing the trunk some distance down from the growing tip of the palm, whereas in a cabbage palm, the fronds drop off more readily, leaving the trunk bare or with only leaf bases attached.

The fruits of *Sabal*, when fully mature and almost dry, are about ¼ inch in diameter; those of *Washingtonia* are larger, about ½ inch in diameter when they are fully mature. Both plants produce large numbers of fruits.

SALIX. There are over 300 species in this group, all called willows. One or more of these species may be found in most parts of the temperate and arctic zones, in swamps in warm climates to just at tree-line in the high mountains. Though most willows are native and wild, some have been used in cultivation, especially the weeping willow, *Salix babylonica*. It is interesting to note that the willows, used by the Indians for medicine, pointed the way to the discovery of aspirin. The young twigs and bark were chewed to relieve headaches, and when pharmacognocists finally analyzed the plant material to discover the active principle, the extracted substance was named salicilic acid, the root of the word salicilic being derived from the scientific name for the genus. This is another example of the many valuable contributions of plants to the well-being of humans, for many need aspirin now and then. The willows produce pollen on one tree and the fruits and seeds on another. Very large numbers of the tiny fruits, with downy hairs attached, blow about in the spring and can be seen floating on air currents long distances away from the nearest willow tree. Cultivation of the trees is quite simple: they prefer, but do not absolutely require, moist habitats. The plants can be grown in full sun but prefer some shade. Most of the species are fast growers and, for this reason, many people like to put them into yards of new houses that have no vegetation on the lots.

————*Salix babylonica,* WEEPING WILLOW. It is not certain where this species is native, but some authorities suggest China. The weeping willow is widely cultivated and grows rapidly to about 30 feet tall. The drooping branches are most attractive, wherever planted, but if you happen to be fortunate enough to have a pond or lake on your property, you can add much beauty to the area by planting one of these handsome plants nearby.

————*Salix nigra,* BLACK WILLOW (fig. 146). This native species is described here not because it might be planted, but because it is so frequent in low, swampy areas where it may grow 40–50 feet tall. The leaves are, of course, deciduous (as are those of all willows) and

Figure 146. Salix nigra (Black Willow)

some of them turn a nice yellow in the fall. So, if one of these trees is growing in your yard, keep it, because it will add much to your landscape. If planting one for decorative purposes, be sure to get a male (pollen-bearing) plant so that you won't have bushels of fine fuzz from the hairy seeds all around.

❋

SAPIUM. There is no one common name for all the species of this genus of the spurge family, the Euphorbiaceae. There are many species, some of which are used for a variety of products. Most of the species come from South America, but the only one from China, *Sapium sebiferum,* is the one we wish to discuss here. None of the members of the spurge family have conspicuous flowers but some of them, such as the poinsettia, have brightly colored leaves surrounding the flowers. Other species that belong to the same family and bear brightly colored leaves that provide outstanding shows are the Codiaeums, commonly called "crotons."

——————*Sapium sebiferum,* POPCORN TREE, CHINESE TALLOW TREE (fig. 147). As one of the common names, Chinese tallow tree, indicates, this species comes from southeastern China where the seeds are boiled to get a waxy substance that can be used to make candles. The seeds may also be crushed to get a drying oil used in paints. But the main use of the tree in the Southeast, where it grows to be 30–40 feet tall, is for decorative purposes. The deciduous leaves are light green in spring and summer and turn bright red or yellow in the fall, thus providing a colorful autumn spectacle. The light gray trunks of the tree are quite handsome as well. The reason that the plant is called the "popcorn tree" is that the bright, white clusters of seeds hang on to the tree after the leaves are shed in the fall, giving the appearance of a plant decorated with popcorn. This plant is said to have been first introduced by Benjamin Franklin when he was the United States envoy in London. It was first grown at Wormsloe Plantation near Savannah about 1785. It was no doubt introduced accidentally several times, and it may have been from one such introduction that the plants became established in Houston, for there are many records of plants in southeast Texas around the turn of

Figure 147. Sapium sebiferum (Popcorn Tree)

this century. They may have been introduced by some enterprising agriculturist who was looking for an additional crop for that region, or they may have come in accidentally in ship's ballast. Whatever, the plants either found the climate to their liking and became naturalized or else were planted by people who found them useful for ornamental purposes. The plants spread eastward and now are found growing throughout regions of the Deep South. We do not recall seeing them in our West Florida neighborhood until well after World War II. They are easy to cultivate, and usually propagate themselves either from seeds or from root runners. There is no need to buy one from the nursery—any person who has one probably will be glad to give you a start from his own yard!

❁

SASSAFRAS. This is another example in which the scientific name and the common name are the same—everybody knows the name, even if they don't know the plants! There are only three species in the genus, the Zone 8 native and two others in eastern Asia. The American sassafras grows over quite a wide range in the eastern half of the United States; it is a smallish tree but is said to grow up to 60 feet tall. We have never seen such a large plant, and most of the ones we know about aren't over 30 feet tall. Most plant parts of all of the species have aromatic oils. The bark and the roots are sometimes harvested to make an extract, sassafras tea, which is now known as a pleasant drink but which used to be thought of as a tonic "good for what ails you." Also, the ground-up parts of the root are called "filé" in Cajun country, a necessary ingredient for making good gumbo.

————*Sassafras albidum,* SASSAFRAS (fig. 148). The trees grow under shade or may be found infrequently in open country. They grow very rapidly when young and the bark is green for some years before the tree matures. The leaves have interesting shapes: broadly elliptic, with one broad lobe near the top (reminding one of a child's mitten) or with two equal lobes near the apex of the leaf. The trees are deciduous and have beautiful colors in the fall, either bright yellow or reddish. Cultivation is very easy: they manage quite well with no attention. Any neighbor who has a sassafras can supply you with a seedling or a root cutting with no trouble at all.

❁

SESBANIA. This genus is a member of the *Leguminosae,* a very large family of plants, most of which produce a fruit called a legume. This is a dry pod with seeds arranged in a line along the length of the pod. In one section of the family, the flowers have shapes somewhat reminiscent of butterflies, described as "papilionaceous." The genus *Sesbania* has about fifty species, some of which are native to parts of the United States, others to Asia or South America.

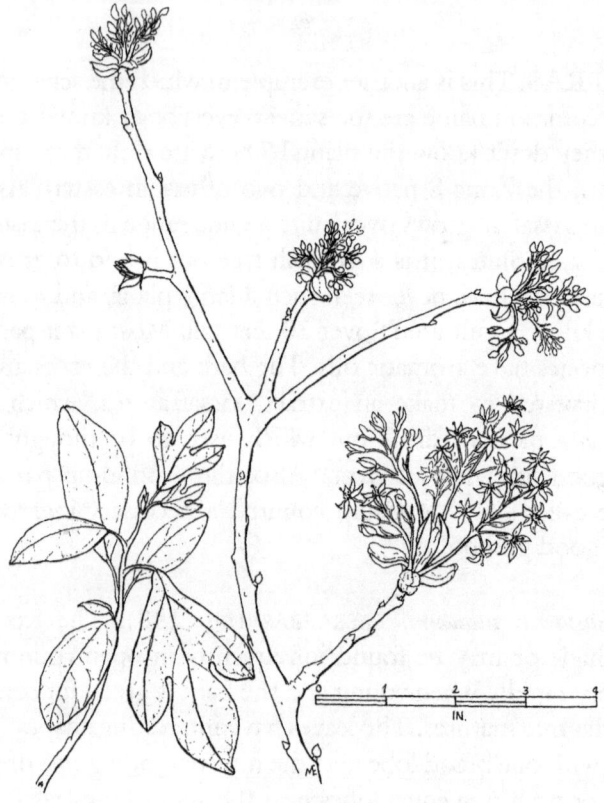

Figure 148. Sassafras albidum (Sassafras)

————*Sesbania punicea,* SCARLET WISTERIA TREE (fig. 149). This species, known mostly as a shrub of 6–8 feet with an umbrella-shaped crown (when grown in an open spot), has become naturalized in Zone 8 from southern Brazil and northern Argentina. Its large numbers of scarlet to bright orange-red flowers arranged in drooping racemes make the plant stand out sharply from its surroundings. The plants have found their way into a number of gardens in this area and flowers just a little after the great burst of bloom in springtime. This species is easily propagated from seed and needs little attention once established. The leaves are compound with 6–12 pairs of leaflets per leaf, their shape providing a nice open, airy look to the plants.

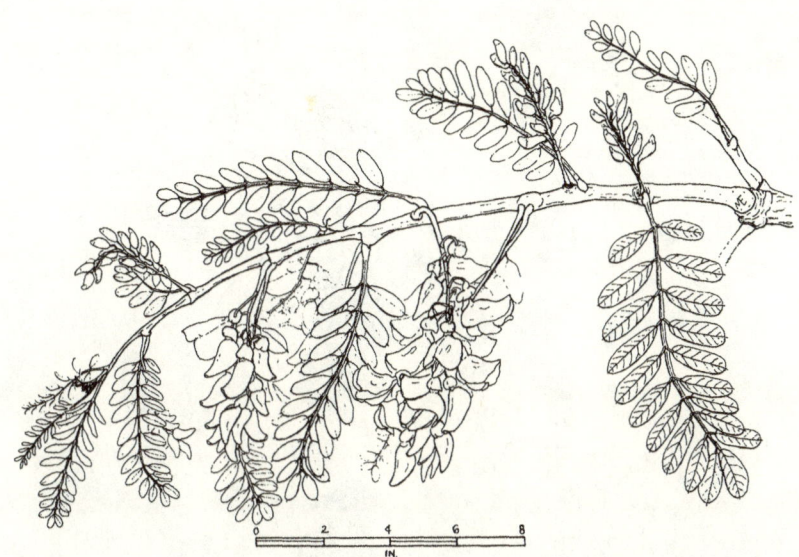

Figure 149. Sesbania punicea (Scarlet Wisteria Tree)

❊

SPIRAEA. All the species in this genus (more than 100, with many cultivated varieties) are called either spirea or bridal wreath. Most of the ones raised in the United States came from China, but some are native to North America. Most of the species are shrubs with small, simple leaves and (usually) wiry branches. The flowers are individually very small, seldom more than ¼ inch across, but they are frequently clustered together in umbrella-shaped clusters (corymbs) to make a very good show. The flowers are most often white, but some are pink and some cultivated varieties bear both colors. Other species may have double forms, as is the case with at least one of the four species that occur in Zone 8. Cultivation is not difficult; a moderately good soil is required, some humus, and fertilizing twice a year for best growth and bloom.

———*Spiraea cantoniensis*, BRIDAL WREATH (fig. 150). This is a double-flowered, white form that flowers from March to April and is one of the more spectacular types growing 7–8 feet tall. The

Figure 150. Spiraea cantoniensis (Bridal Wreath)

shrub has a pleasing, rounded shape, with branches from the ground up and all around. This growth habit is common to many of the species; of course, they achieve this form only if they're not crowded or too shaded.

―――――*Spiraea prunifolia,* BRIDAL WREATH (fig. 151). This shrub is a bit smaller than the former, and most frequently has double white flowers as above. It may flower a bit earlier.

―――――*Spiraea thunbergii* (fig. 152). *S. thunbergii* flowers very early (late February–early March), making a spectacular show of tiny white clusters on shrubs 7–8 feet tall. Some people call this shrub "baby's breath," but that name is most frequently reserved for a quite different herbaceous plant, *Gypsophila paniculata*.

Figure 151. Spiraea prunifolia (Bridal Wreath)

———*Spiraea* × *bumalda* (fig. 153). This bridal wreath usually flowers later than the two above, sometimes flowering through the summer. It may have pink and/or white flowers in corymbs that are somewhat larger (1–1½ inches across) than the corymbs of those species described above.

❀

STEWARTIA (sometimes spelled *Stuartia*). There are only about 6 species of this genus in the world, some in eastern Asia and only one known from Florida, the one following. This genus is in the same family as the camellias, and indeed, members of the two genera resemble each other. Species in *Stewartia* are all single flowered and have deciduous leaves, whereas the camellias are evergreen and frequently double flowered.

Figure 152. Spiraea thunbergii (Bridal Wreath)

―――――*Stewartia malacodendron,* SILKY CAMELLIA (fig. 154). These plants have a very restricted range in their native habitat and have been placed on an official list of endangered species. Therefore, it is unlawful to collect these plants from the wild. Fortunately, you can find a few nurseries who propagate the species, so that you can get a specimen if you want it badly enough. We have only seen a few plants growing. The flowers are quite large, 3–4 inches across, crinkly white, with large numbers of purple stamens. The plants may grow 8–10 feet tall.

❦

STYRAX. This genus and another in the same family, *Halesia* or the silver bell tree, provide some very nice, early spring–flowering

Figure 153. Spiraea × *bumalda*(Bridal Wreath)

plants. Some species are from eastern Asia, others from eastern North America, and at least one grows in southeastern Europe. There are both small trees and shrubs in the genus, and several of them are in cultivation. *Styrax benzoin* is the source of the medication benzoin.

———*Styrax americanus,* SNOWBELL (fig. 155). Snowbells are shrubby, not more than 10 feet tall. They grow well in shady areas, particularly well under pine trees. The deciduous leaves are ovate and light, dull green, 2–3 inches long. They have barely perceptible wavy margins. The white flowers are carried singly at almost every node (joint) of almost all of the branches. They hang down, much like flowers of the silver-bell tree, are usually five-lobed but occa-

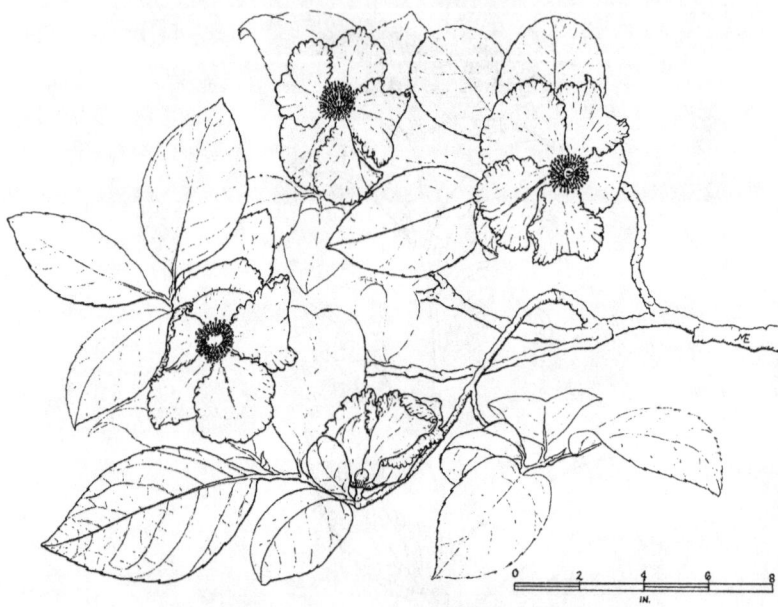

Figure 154. Stewartia malacodendron (Silky Camellia)

sionally four-lobed, and the whole flower is about ¾ inch across. The flowers are followed by small, round, oval fruits. As natives, these plants are well adapted to our sandy soils.

❁

TABERNAEMONTANA. This genus is mostly tropical, and the plants of interest in Zone 8 are quite sensitive to cold weather. If you like the challenge of providing a location that can be kept more or less frost free or are willing to have a plant that will be killed back by a freeze but that comes back up each spring, then this is an interesting group.

————*Tabernaemontana divaricata,* CRAPE JASMINE, CRAPE GARDENIA, PINWHEEL FLOWER, and several other names. This is a much-branched shrub with white flowers that are fragrant at night. The ones we have seen seldom get more than 3–4 feet tall because they die back with any freeze and the next year's growth doesn't have a chance to get any taller. But if the plants are in a frost-free

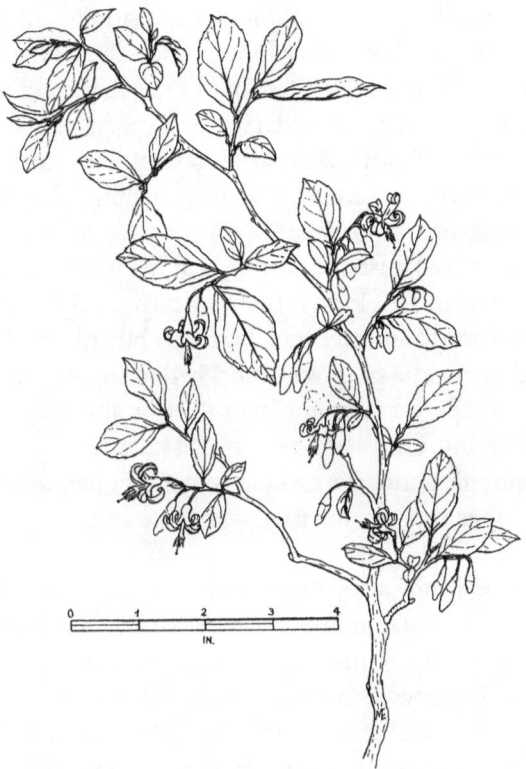

Figure 155. Styrax americanus (Snowbell)

environment, they may grow to 8 feet tall. The leaves are opposite each other on the stems, somewhat variable in shape from oblong to lance-shaped to obovate (wider at the end than at the base), one of the opposing leaves usually larger than the other. The white flowers are waxy, about 1½ inches across, somewhat fragrant. Plants may be grown from seed or from cuttings.

TAXODIUM. This genus has only three species from North America, two in the eastern United States and the other in Mexico. These handsome trees, known commonly as cypress, share that common name with many other species not at all related (or at least not closely related). In their native habitat the trees grow either on the shores of rivers, lakes, and swamps or actually in the water. Most Southerners are familiar with cypress "knees," those woody projections that rise as much as 5–6 feet out of the water and act as devices to take oxygen to the roots underwater. Cypress lumber is very valuable because it will not rot under water and because it has a beautiful grain valuable in furniture or in panelling. The trees are less frequently regarded as ornamentals to be used in our yards, sometimes because of the mistaken notion that the plants must grow with their feet in water. In Florida, one can buy young plants from the Forestry Division, and it is much better to buy plants than to try to dig up wild ones. The trees very quickly develop long tap roots that are hard to take intact from the wild. Just be sure to give the plants frequent watering until they have had a chance to get those roots down far enough under the ground to get to a permanent supply of water. After that they grow easily, with little trouble.

———*Taxodium ascendens*, POND CYPRESS (fig. 156). This species is, as the common name implies, most often found in ponds and bogs. It is generally a smaller tree than the following species, the bald cypress. The species name *ascendens* refers to a characteristic of the needle-like leaves of the pond cypress which, on the younger twigs, ascend or stand up from the twig on two sides so that the

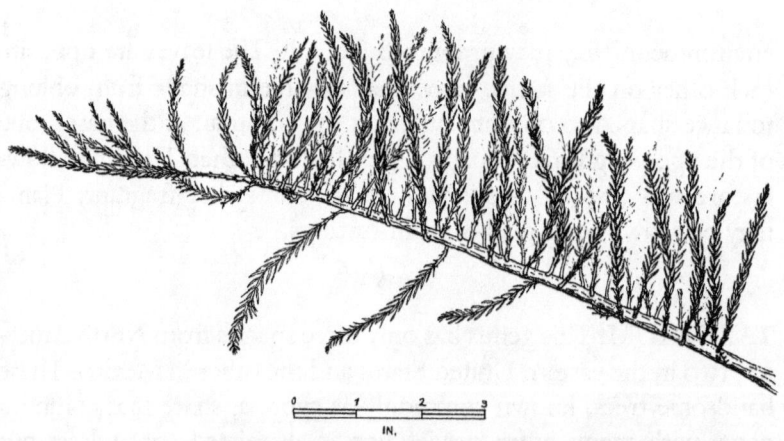

Figure 156. Taxodium ascendens (Pond Cypress)

leaves appear as V-shaped structures when viewed from the tip of the twig. In contrast, the leaves of the bald cypress stand more horizontally from the midrib. The bald cypress leaves have a more "feathery" look than do those of the pond cypress. This species is not commonly propagated by the Forestry Division, but the following species is.

———*Taxodium distichum*, BALD CYPRESS (fig. 157). The term "bald" implies that this tree is deciduous, like both of the other species in the genus. This is one of the few cases of deciduous leaves among the conifer group, whose members, in general (such as pines and junipers), are evergreen. The bald cypress grows to become a

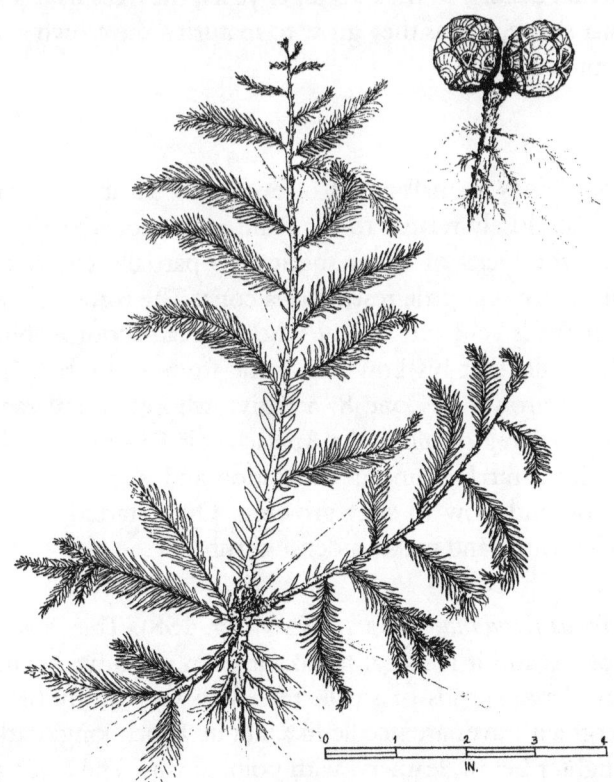

Figure 157. Taxodium distichum (Bald Cypress)

majestic tree up to 150 feet tall. The natural habitat of this species is not too different from that of the pond cypress described above; it frequently lives with "its feet" in water but sometimes grows up on the margins of streams, rivers, or fresh water lakes. As indicated above, the leaves of the bald cypress have a more horizontal, feathery appearance than do those of the pond cypress. As with other conifers, the seeds are produced in cones, which are round and about 1 inch in diameter. These trees make very fine additions to yards and gardens, both in natural stands and as planted specimens. Bald cypress will grow very well away from low areas. When grown in higher ground, trees usually do not develop the typical knees found with the plants in wet habitats. Bald cypress may be grown in a variety of settings but should be given plenty of room to develop its full beauty. In their younger years, the trees have a slender, triangular shape, but as they grow to maturity, the crown assumes a more rounded shape.

❧

TAXUS. This is a small genus of evergreen shrubs or small trees that is most closely related to the conifers, or cone-bearing plants. However, the seeds of *Taxus* species are partially covered with a fleshy structure that little resembles a cone. The common name for all the species is yew, and anyone who has read about Robin Hood certainly recalls that his bow was made from a yew tree. There is one species grown in Zone 8, a native whose natural range is a relatively small region along the Apalachicola River in west Florida. This species is rarely found in cultivation and is said to be difficult to establish and slow to start growing. Once started, however, it grows vigorously and makes a dense shrub.

———*Taxus floridana,* FLORIDA YEW (fig. 158). This is an endangered species and it is illegal to take it from its native habitat. The plants are large shrubs or small trees with very bushy branching. The evergreen leaves are needle-like, about 1 inch long, dark green above, lighter below, leathery, with pointed tips. The seeds are half enclosed in a red, fleshy structure, the aril. Unless you have a green thumb and plenty of patience, it would be best to substitute another

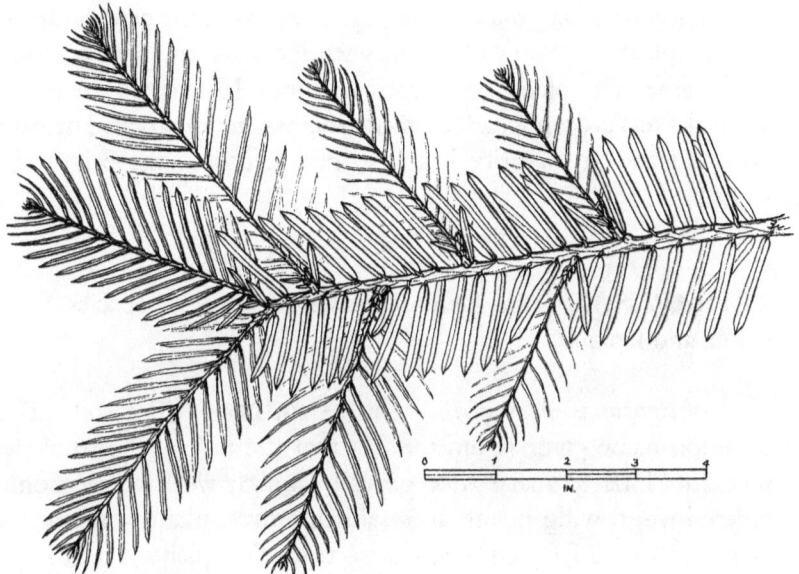

Figure 158. Taxus floridana (Florida Yew)

plant for this one. However, some people like the challenge of raising "difficult" plants and this one would certainly be rewarding if cultivation is successful. Considering all the fuss it requires, you should use it in as near its natural form (a rounded bush) as possible, as an accent plant. It can be trimmed or pruned if necessary. We have not seen any record to indicate whether the Florida yew is, or is not, poisonous. However, it would be wise to treat it as poisonous because all the other species of *Taxus* are.

TERNSTROEMIA. In a number of publications, the name *Cleyera japonicum* is erroneously used by some nurseries for the species of *Ternstroemia* described here. *Cleyera japonicum* is yet another plant, rarely grown in Zone 8. *Ternstroemia* is another group that may not be familiar, but the plants are interesting and quite handsome. They are members of the same family as the camellias (Theaceae) and their culture is quite similar. They are evergreen with glossy, deep green leaves. The plants may be propagated by cuttings, transplant easily, and grow well in semi-shaded areas.

———*Ternstroemia gymnanthera* (fig. 159). As in the case of camellias, the plants are known by their scientific name and have no common name. The plants are shrubs to about 15 feet tall, or more. Their flowers are white, about ¾ inch across, and fragrant. The fruit is round and at maturity it splits open, exposing the bright red seeds.

❦

TETRAPANAX. This genus consists of but one species from China and Japan.

———*Tetrapanax papyriferus,* RICE PAPER PLANT (fig. 160). The common name comes from the fact that the pith of the trunks is used in China to make "rice paper." Though we have seen only rather low-growing plants, it is said that these plants can reach a height of 10–12 feet. The major attraction of the plants is their large leaves, 10–12 inches across, with deep cuts in the margins. But this

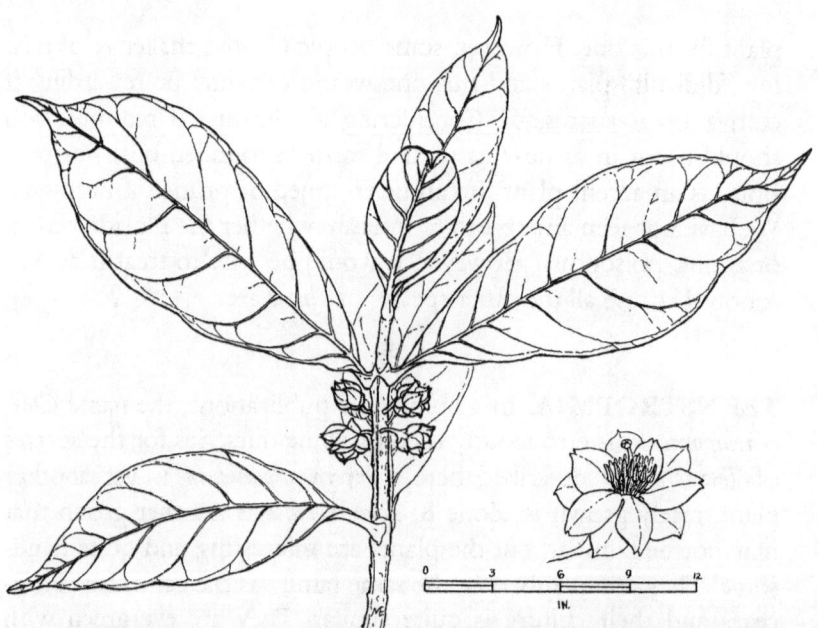

Figure 159. Ternstroemia gymnanthera. (after Beddome, *Flora Sylvatica for Southern India,* vol. 1, Table 91, 1871)

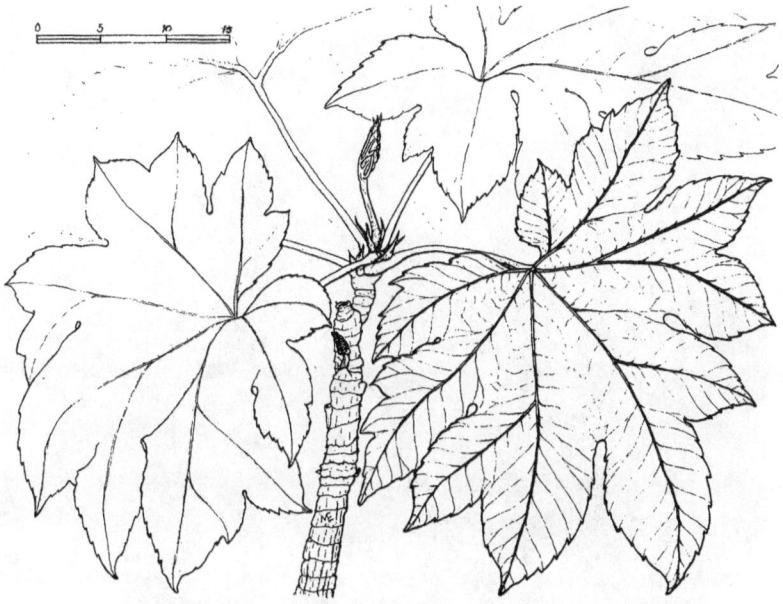

Figure 160. Tetrapanax papyriferus (Rice Paper Plant)

species has the added attraction of the white flowers that project above the leaves in December. The plants grow well in shade and the most successful ones are found on the north side of houses. The plants are described as "root hardy" because the roots will survive while the leaves may be killed back by cold weather.

THUJA. All of the five species of this genus are called arbor vitae. A coniferous group, all are evergreen and occur naturally in North America and eastern Asia. They are easily distinguished from other cone-bearing plants (such as pines and cypresses) by the manner in which the smaller branches are arranged: they appear as flattened, erect fans, many occurring close to one another all around the tree. Most of the *Thuja* species are trees, but there are many that are shrub-like.

———*Thuja occidentalis,* ARBOR VITAE, WHITE CEDAR (fig. 161). This arbor vitae is found in eastern North America, from Nova Scotia south to North Carolina. It is a tree, up to 60 feet tall but, like

Figure 161. Thuja occidentalis (Arbor Vitae). Plant 10 ft. tall.

all the other species, has many cultivated varieties that are shrubby. Obviously its native range has been much expanded as there are many cultivars in the Deep South. One may choose variations that have yellowish green or whitish tips to the foliage. Some are referred to as "golden," with bright yellow foliage. Cultivation is easy, with just the ordinary amount of care, watering and fertilizing. These plants make excellent hedges or may be used as accent plants at corners of the house.

TRACHELOSPERMUM. There are about ten species of this genus with its jawbreaker name, most in tropical and subtropical Asia, and a deciduous one in North America. All the other species are evergreen vines, climbing up supports and trees 50–60 feet under normal conditions. However, because of these plants' subtropical nature, they are liable to be killed when Zone 8 has a colder than average winter. The plants are readily propagated by cuttings taken in spring or early summer.

———*Trachelospermum jasminoides,* CONFEDERATE JASMINE, STAR JASMINE (fig. 162). It has been sad to see very old and beautiful vines of the Confederate jasmine killed by the extra cold weather of the late 1980s. This vine, grown by many gardeners, is particularly nice to have because the numerous white flowers in spring and early summer give off a delicious fragrance that permeates the air for some distance. We are on our third try to grow the plants from cuttings—this time they may make it although the first two were

Figure 162. Trachelospermum jasminoides (Confederate Jasmine)

killed by the cold. Why this one escaped a recent winter's very cold weather, we don't know. We just hope the plant will continue to thrive. The vines will grow quite dense on trellises or will twine about the trunks of any tree they're planted near, so you can use them in a variety of settings.

❦

TRACHYCARPUS. This genus is in the palm family, a family well-represented in the tropics, with only a few genera in the south temperate region. All species of *Trachycarpus* originated in areas of southeastern Asia, including southern China, Burma, and countries to the east and south. The species in this genus generally have the male and female parts on separate plants; none of them have both sexes in the same flower. The following species is the most prominent one of the genus, and is widely planted in the Deep South.

————*Trachycarpus fortunei,* WINDMILL PALM (fig. 163 a and b). The species is named for a famous explorer in China and Japan, Robert Fortune, who lived from 1812 to 1890. This is a fan palm (leaves broad and with one expanse) but the segments of the leaf may either hang down or remain essentially horizontal. This apparently causes the leaf to move in somewhat circular directions in the wind, hence the name "windmill palm." This palm is easily distinguished from other palms in Zone 8 by the thick, dark brown to black coat of hairs surrounding the stem below the green leaves (see fig. 163 b). These hairs are attached to the sides of the leaf bases and hang on for some time, on older trunks occurring as much as 2–3 feet below the leaves. As with other palms, the plants should be considered as accent plants and given sufficient room to develop. This is almost as hardy a species as the needle palm, *Rhapidophyllum hystrix.*

❦

TSUGA. The hemlocks are natives of the northern parts of the eastern and western United States, and the species mentioned below occurs as far south as the mountains of North Carolina. The hemlocks are evergreen, cone-bearing plants with numerous cones about

Figure 163 a and b. Trachycarpus fortunei (Windmill Palm). Plant 10 ft. tall. 163a–Foliage. 163b–Brown-black hair on trunk.

1 inch long. The trees are very decorative and are used quite often on large estates. The dark green foliage is rather feathery and makes handsome patterns of sunlight and shade. The plants may be treated as hedge plants, and take to pruning quite well. Local nurseries may not stock the species that will grow here; however, many times people have brought small plants home in their car. If they are good gardeners, they may succeed in keeping them going. A tree 25 feet tall is growing well in our town.

———*Tsuga caroliniana,* EASTERN HEMLOCK (fig. 164). These plants are restricted to the mountains of southwest Virginia and North Carolina. They are handsome components of the native vegetation of this area. The tree typically grows into a narrow, pyramidal shape, but sometimes more than one trunk develops and this may change the overall appearance of the mature tree. The evergreen leaves are dark green at maturity but, like many pines, put on new leaves that are much lighter than the mature ones further back on the branches, making quite a handsome display in the late spring. As mentioned above, these are cone-bearing plants, with small elliptical cones producing numerous small seeds. The seeds probably

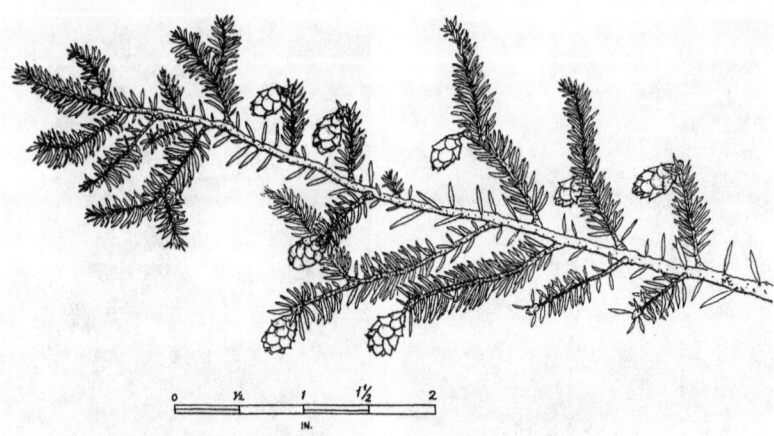

Figure 164. Tsuga caroliniana (Eastern Hemlock)

would be difficult to propagate in Zone 8. The tree will need a good amount of space and should not be crowded so that it can attain its very symmetrical shape.

❧

VACCINIUM. Plants in this large genus are best known for their small fruits, such as blueberries and cranberries, rather than for their decorative or ornamental value. There are other genera in the same family, Ericaceae, that include very beautiful ornamentals, such as the rhododendrons and azaleas. There are also some native members of the genus *Vaccinium* that have some merit in ornamental planting, such as the following.

————*Vaccinium arboreum,* SPARKLEBERRY (fig. 165). The sparkleberry is a medium to tall shrub, growing as tall as 15 feet. It is generally evergreen, though the leaves tend to get rather scarce in the late winter. The leaves are small and oval, bright, shiny green above, about 1 inch long. In April–May there are many small, white to pinkish, bell- or urn-shaped flowers that hang with the open end down. The flowers, which attract honey bees, are followed by small, round, black berries that are very shiny (thus the name sparkleberry). The berries also attract many birds. The shrub has an overall umbrella shape and the many stems are rather attractive themselves.

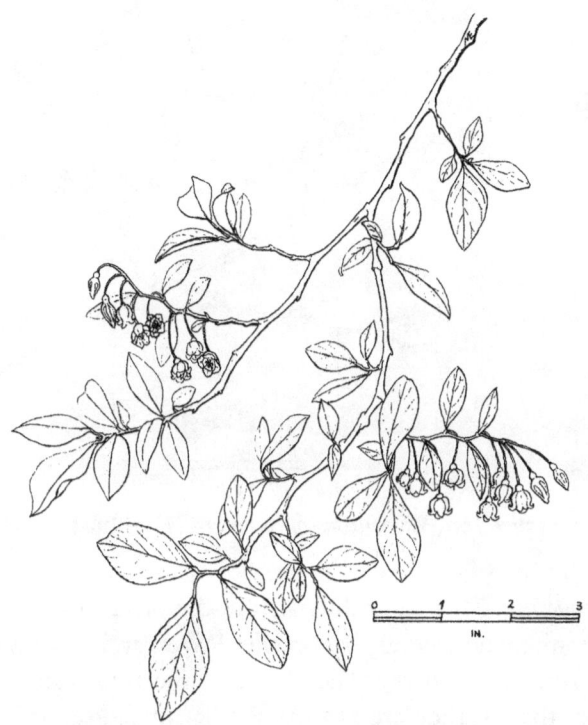

Figure 165. Vaccinium arboreum (Sparkleberry)

Unless your yard is very crowded, this shrub would make an excellent addition to a native border. Such a native border might include sparkleberry, yaupon, beautyberry, and even a wild grape such as muscadine growing up through the shrubs to provide both protection and food for birds. Cultivation is not difficult—practically no care is needed after the shrub is planted. You can probably find a friend who would be glad to give you a start.

———*Vaccinium corymbosum,* HIGHBUSH BLUEBERRY, HUCKLEBERRY, SWAMP BLUEBERRY, WHORTLEBERRY (fig. 166). This species has been given several different scientific names, such as *Vaccinium fuscatum, V. atrococcus,* and others, but on advice from a specialist, it is better to lump them all together under one name, *V.*

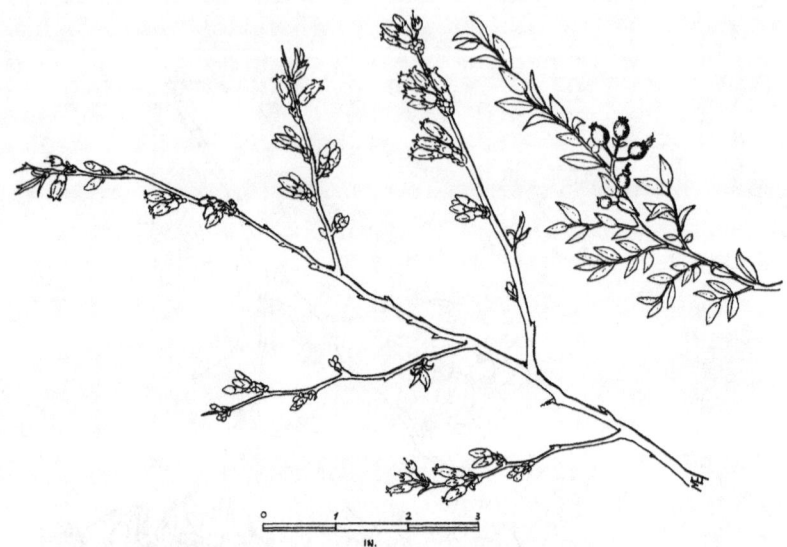

Figure 166. Vaccinium corymbosum (Highbush Blueberry)

corymbosum. This plant differs from the preceding species in that it is definitely deciduous, has smaller leaves, and blooms much earlier than the sparkleberry. The shape of the two plants is nearly the same; that is, they are both umbrella- or fan-shaped and they are about the same height (12–18 feet or sometimes more). The two species grow in the same habitat, shady pine-oak woods, and have fruits that are nearly the same size and color. Both have edible fruits, much sought after by birds. The highbush blueberry has been used in hybrid programs to produce many cultivars used for fruits. Its horticultural value lies in the light green color of its early leaves; later, in the fall, its leaves add a pleasant, brownish tone to the dying foliage. Also, the irregular branching of the limbs and the very numerous branchlets give the plants an interesting visual texture.

VITIS. This is the genus of grapes, both the native and the cultivated species from Europe and Asia. Following our own research, there are approximately 28 species of grapes native to the North

American continent. No wonder the Vikings (who, many claim, were the original discoverers of our continent) called the land they found "Vinland." Many species are referred to as fox grapes, but that name must be reserved for one that is common in the eastern United States (though infrequent in Zone 8b), *Vitis vulpina*. Grapes native to the Southeast are either "bunch" grapes or muscadines. Most species are in the former group, and only two ill-defined species are included in this reference as muscadines. Though most of the grapes are thought of only in terms of their fruit-bearing capacity, the vines themselves can make handsome, decorative, horticultural specimens. There is no more pleasant experience on a warm summer afternoon than to sit in the shade of an arbor of scuppernong grapes, particularly in mid- to late August and early September when the grapes are ripe. Then one can enjoy the plants for their shade and the fruit for their great flavor. This implies, of course, a revival of this type of culture of the grape, through which it is updated to the status of a "patio shade" plant. Or, the plants can be used as a sun screen for sunny windows; because they are deciduous, they do not obstruct the sun in winter.

————*Vitis rotundifolia*, MUSCADINES, SCUPPERNONGS (fig. 167). There are two main types of muscadines, each easily identifiable by the color of the mature berries. There are many varieties of tan to light green muscadines as well as many of the dark purple types. In the former group, the scuppernong is no doubt the best known, but there are many tan varieties developed later that have larger or sweeter fruits and it is becoming harder and harder to find the original scuppernong. The grape vines grow very well and rapidly, and after about three years, they should begin to bear fruit. Since these vines are trained on to trellises or overhead arbors, one must establish which type of training one wants at the time of planting. After the vines have achieved the size desired, they should be rather ruthlessly pruned each winter, no later than January. Most people are afraid to take off enough vine for fear of killing the plant. It is practically impossible to kill the plants unless one digs them up roots and all or hits them with a weed killer. As mentioned earlier, the plants can

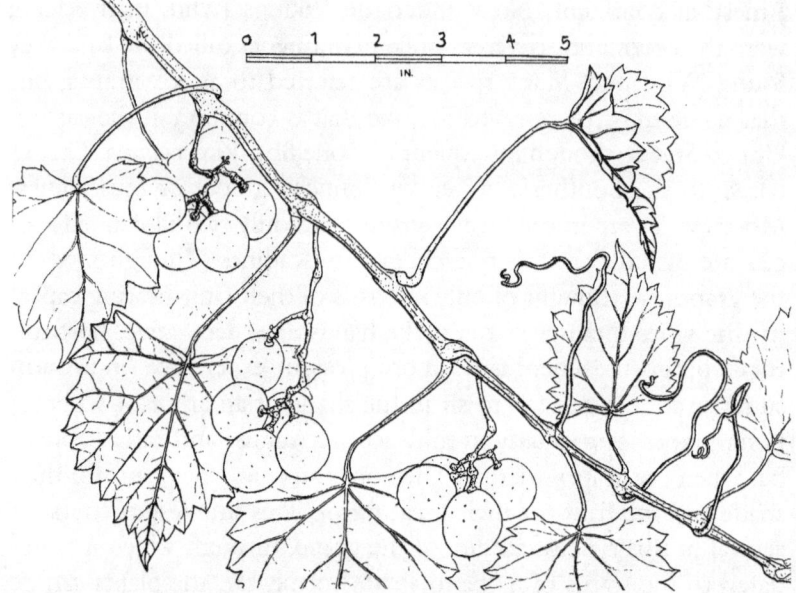

Figure 167. Vitis rotundifolia (Muscadine Grape)

be trained in many different ways and lend themselves almost to the kind of sculpturing one finds with boxwood. Thus, they can serve a double purpose, delight of the eye and the palate!

❦

WASHINGTONIA. There are only two species in this genus of palms. They occur from the dry interior regions (but near springs) of southern Arizona and California south into the Mexican states of Sonora and Baja California. Both species are magnificent members of the palm family and are certainly fitting tributes to George Washington, for whom the genus is named. Both species are bisexual, having male and female organs in the same flower.

———*Washingtonia filifera*, WASHINGTONIA, (fig. 168 a and b). This is the more northern species, and is better adapted to the Zone 8 climate than the Mexican species, *W. robusta*. It grows 50–60 feet tall (75 feet maximum). The fronds are larger, a brighter green, and

Figure 168. Washingtonia filifera (Washington Palm). 168a–Plant up to 50 ft. tall. 168b–Dead leaf still attached to trunk. Sharp teeth are faintly visible on leaf stalk.

hang on to the trunk longer than those of the cabbage palm. The fan is nearly flat, and the petiole (leaf stalk) is armed with rather stout teeth. The plants also seem better able to withstand the cold weather than the cabbage palm, which is native to the regions from about Panama City, Florida, south and east and up the Atlantic coast to the coastal islands off North Carolina.

Washingtonia resembles *Sabal palmetto* superficially. Please see the distinctions given under the discussion of *Sabal palmetto*.

❈

WEIGELA. This is a genus of 10 or 12 species, all from China and Japan, that are more common in northern gardens than in those of Zone 8, though there are a few individuals planted in this zone. At any rate, one should consult a local nursery to determine if it recommends this genus. There may have been some developments that make these plants more amenable to our climate. If such is the case, it would be worthwhile to have one or two of these handsome, deciduous shrubs.

————*Weigela florida* (fig. 169). With no known common name, plants in this species are generally called by the generic term alone. There are two possible species that could be used here, but the one listed here is the only one we have seen. The other species is *Weigela floribunda*, with scarlet flowers. *W. florida* has pink flowers, sometimes with some white on the corolla. Note that the species name is *florida*, but in this context it means "flowering" and not the state of Florida. These are medium-height shrubs (8–10 feet), deciduous,

Figure 169. Weigela florida (Weigela)

with rather dense clusters of flowers near the tips of the branches. We have seen only one or two plants growing in Zone 8, and thus do not have an accurate idea of their real capability here. Those spotted were rather less vigorous than those from farther north, but there are all kinds of possible reasons for this, not just the wrong habitat.

❦

WISTERIA. The name of this genus honors Caspar Wistar (1761–1818), Professor of Anatomy, University of Pennsylvania, but the genus was spelled *Wisteria* when its description was published. There are 9 or 10 species of *Wisteria* in eastern Asia and eastern North America. This group of vines belongs to the bean family, the Leguminosae. There are probably several hybrids among them, but we describe only one here. Vines are very interesting members of the plant kingdom because of their various means of hanging on to other vegetation as they grow upward: some attach themselves by tendrils, others use modified "suction cups" to grab on to some support, and others twine about the stems of the plants they grow on. This does not make them parasites, though their vigor may cause the supporting plants to become choked off from light. It seems that all vines are "heliophiles"—sun seekers—so they grow as rapidly as they can to the top of whatever supports them. The wisterias are twiners, and may grow up 50 feet or more.

────*Wisteria floribunda,* JAPANESE WISTERIA, WISTERIA. Several varieties of this species are used in cultivation all over the temperate zones on most continents. Some of the variations have purple to lavender flowers, others white. Both of these types are found in Zone 8, but purple predominates. Many plants have escaped and become naturalized, and at many old house sites wisteria remains as one of the last vestiges of former occupation. After the beautiful descending or hanging flower clusters have dropped, there frequently remains a velvety pod with enclosed seeds. The seeds are but one means of propagation; the stems running along the ground root readily at intervals. One may artificially take advantage of this latter trait by a process of layering—covering a section of the stem still attached to the parent plant to produce roots, then after several weeks or months, severing the stem with the attached roots for a new plant. The wisteria's propensity for propagation makes it something of a weed, but such a beautiful weed! If one has the patience, one may train the vines into self-supporting small "trees" by careful and judicious pruning.

———*Wisteria sinensis* (fig. 170) is also commonly grown. It is hard to distinguish from the former species. There is a white form, *alba,* found frequently.

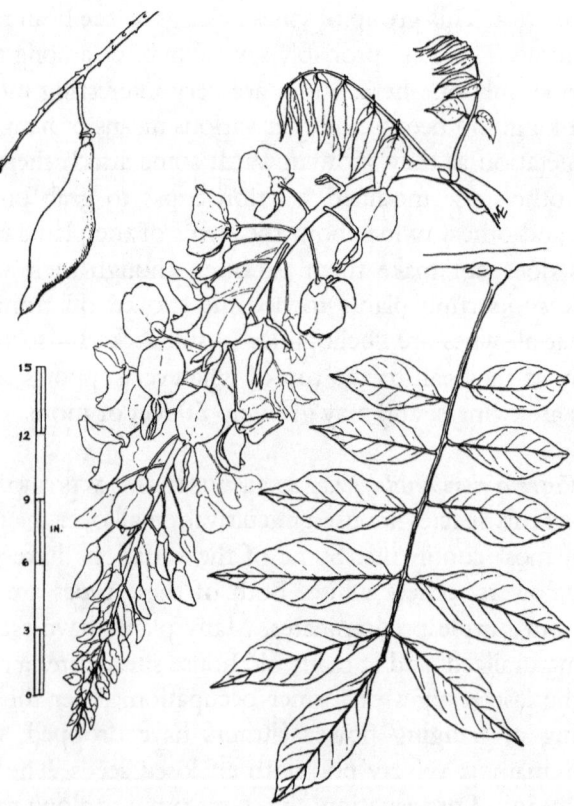

Figure 170. Wisteria sinensis (Wisteria)

❀

YUCCA. The yuccas are generally associated with the desert of the American Southwest, but a few species of this genus occur in the Southeast. Their habitats are generally coastal and inland deep sand areas, which are, in a sense, like deserts because whatever moisture falls passes rapidly through the upper 8–10 feet of soil without much benefit to shallow-rooted plants. Most of the species are rather short-stemmed, not much over 10 feet tall, sometimes branched but more frequently with a single stem. The leaves form a dense cluster around the stem, from the tip to the ground. They are all strap-shaped, some leathery and stiff, others more flexible. The species all produce many flowers from the tip of the stem, from white to greenish white to yellowish to red. Usually several large fruits (dry capsules) are found on each inflorescence, with many seeds in each.

————*Yucca aloifolia,* SPANISH BAYONET (fig. 171). This is the species of yucca most commonly used for decoration in Zone 8. It gets its common name from the very sharp spines at the end of each leaf. The leaves have been used in the past to provide fibers for string and rope. This species is evergreen. The stems may grow 10–12 feet tall and will hang on to all of their leaves, from top to bottom. This native produces very handsome white flowers, clustered closely on the inflorescence just above the leaves. Plants are quite striking in scrub-oak areas, the driest type of habitat in this region. Because of this preference, this species makes very good plantings for those who have dry soils and don't care to keep watering plants, as is necessary for many other ornamentals. One must be prepared for spreading—these plants will increase in number from a single start, and removing large ones is no easy chore. They transplant easily, provided one takes very small plants. They make attractive hedges or fences, protecting against intrusion by dogs or preventing folks from taking short cuts across the lawn.

————*Yucca flaccida,* DEVIL'S DARNING NEEDLE, BEARGRASS (fig. 172). This species grows in much the same type of habitat as does

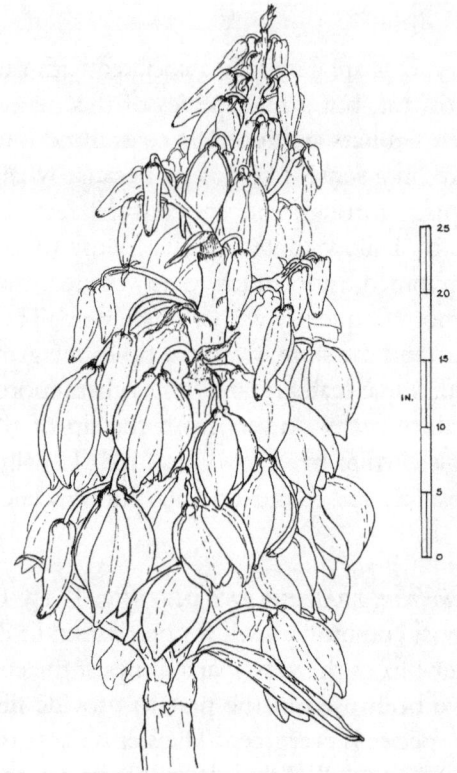

Figure 171. Yucca aloifolia (Spanish Bayonet)

the Spanish bayonet but, rather than having elongated stems, the leaves form a rosette near the ground and never elongate. The leaves are more flexible than are those of the former species, are narrowed at both base and apex, and generally have filamentous hairs scattered along their margins. Like the preceding species, *Y. flaccida* has very sharp, spiny points (thus the name, "devil's darning needle"). If you go out to dig a plant from the wild, be prepared for quite a dig because the plants have long, thick caudexes (underground stems) for some distance down. So take small plants which may be offshoots from larger ones next to them, and even if you don't get all the roots, the plants generally will form new ones if they are kept watered at first. The flowers are produced in loose clusters on bare stalks 3–4 feet above the center of the basal rosette of leaves. Cor-

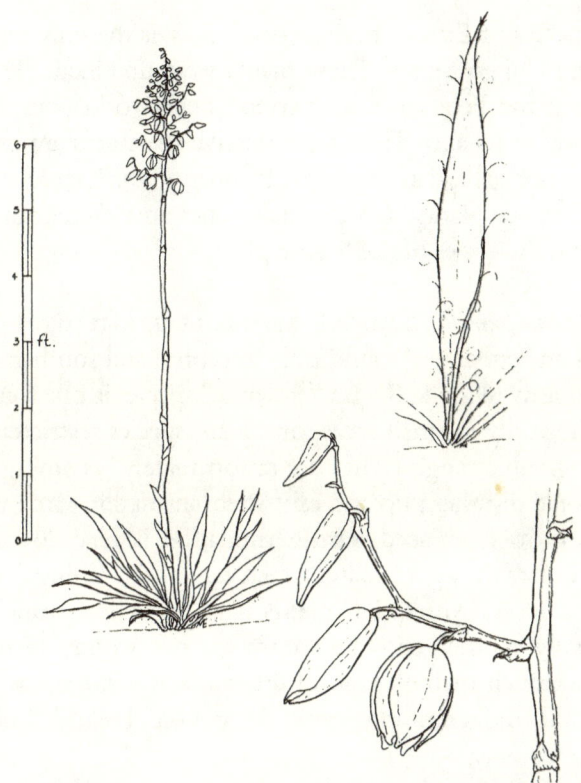

Figure 172. Yucca flaccida (Devil's Darning Needle)

sages have been made from the flowers—but one had better be very careful to get rid of all the ants that love these flowers before giving the corsage to one's girl friend! If you wish to plant bear grass, one should plant several in an area to get the best effect. Put them about 3 feet apart.

❦

ZAMIA. This genus is one of a few "relict" genera of the general group referred to as the cycads. (A "relict" refers to a species or genus that lived and thrived in an early geologic age, and which still grows today, though in reduced numbers or smaller geographic areas.) The only other member of this relict group that we have represented in this book is the sago palm, *Cycas revoluta*. There are several species of *Zamia*, but the one below is the only representative in the United States. These plants were dominant elements in the flora of the Pennsylvanian period of the world's development, 325 million years ago. Their reproductive structures are cones, not fleshy, but not as hard as any of the familiar pines. There are separate male and female plants, and the male cones are different in size and shape from the cones of the female plant.

———*Zamia pumila*, COONTIE, SEMINOLE BREAD (fig. 173). This species is an "endemic," found only in central and southern Florida but fully hardy in Zone 8. (An "endemic" species is one found in its natural area only in small areas, or a plant species restricted to very small geographic range.) They are unfortunately becoming rare because people dig them up and take them into their gardens. While they will grow from seed, seed germination is very slow and one gives up before they germinate, which can take anything from a few months to several years. The plants have very short trunks, sometimes totally submerged. The trunks are not woody in the usual sense, and much of the tissue in the trunk is a storage region for starch, thus the common name "Seminole bread." The leaves

Figure 173. Zamia pumila (Coontie). Plant 3 ft. tall.

(fronds) grow from the apex of the trunk and are compound, with stiff, strap-shaped leaflets. The whole leaf may be from 15–30 inches in overall height, and is evergreen. The male and female reproductive structures emerge from the center of the apex of the trunk, male and female structures occurring on separate plants. The cones are a rich, medium brown, and may be 5–6 inches long, the female about 2 inches in diameter and the male 1–1½ inches in diameter.

❧

ZIZYPHUS. It is difficult to know where to draw the line between ornamental and fruit trees, particularly when the plant under consideration is attractive or has some attractive feature in addition to its fruit. The plants included in this genus with a very interesting name have fruits that are eaten, and there are a number of cultivated varieties with improved fruits. The species are found in various parts of the world, from southern Europe to eastern Asia and in the southwestern United States.

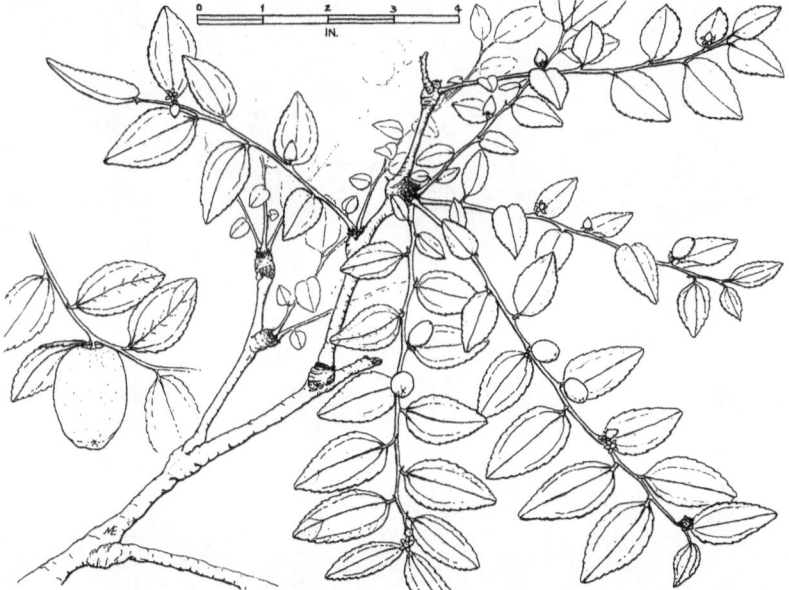

Figure 174. Zizyphus jujuba (Common Jujube)

———*Zizyphus jujuba,* COMMON JUJUBE, CHINESE JUJUBE, CHINESE DATE (fig. 174). As implied by the common names this species, the only one we have encountered, originated in China. It is a large shrub or small tree to 30 feet tall, with deciduous, shiny leaves. If one were to use the plants as ornamentals, it would be for the handsome foliage. The small, white flower grows in clusters in the axil of the leaf and stem. The fruits have about the same structure as either a date or an olive and are about the same size as a small olive (at least in those varieties that have not been improved). The plants grow easily from root sprouts or from seeds and transplant easily. They do well in open areas, with only open shade at most. One caution: there are sharp spines on each side of the base of the leaves. Your children would not climb in it, nor would the deer likely browse it!

Tables of Horticultural Characteristics and Landscape Planning Aids

Tall Trees (90' or taller)

Plant Name[1]	Ht. at Maturity[2]	Leaf Retention[3]	Light Preference[4]	Major Horticultural Features[5]
Bald Cypress *Taxodium distichum*	150'	D	S-PS	Feathery foliage, orange-brown in fall
Camphor Tree *Cinnamomum camphora*	100'	E	S-PS	Rounded crown
China Fir *Cunninghamia lanceolata*	120'	E	I	Tall, pyramidal shape; glossy foliage
Cryptomeria *Cryptomeria japonica*	150'	E	I	Statuesque tree, graceful branching
Deodar *Cedrus deodara*	150'	E	S-PS	Light, gray-green foliage
Longleaf Pine *Pinus palustris*	120'	E	S	Tall, stately tree
Maidenhair Tree *Ginkgo biloba*	120'	D	S-PS	Globe-shaped tree with interestingly shaped leaves turning golden in fall

Please note: This table is arranged alphabetically by common name. None of the characteristics in the table is absolute—the heights given may never be achieved in Zone 8 or plants may reach greater heights than those listed. Light preferences are equally difficult to pinpoint because the individual specimen of a particular plant may be well adapted to more or less shade. The major horticultural features listed here may, or may not, agree with your own observations. Equally, the suggested major use might not agree with your use of the particular plant. All such observations are generalities that may help you choose one plant or another. The heading "undesirable traits" is included because most nursery workers probably will not disclose that a plant you are about to buy has any undesirable characteristics.

1. For each species listed in this table, more information may be found in the text. Use the scientific name to locate any desired plant.
2. The table uses heights of plants as the major division, but we would not argue if you think some of the plants are in the wrong division; the border between small trees

Major Use	Flower Time[6]	Flower Color[7]	Fruit Color[7]	Undesirable Attributes
Specimen	NA	NA	NA	
Shade, trimmed hedge	S	Green	Purple	
Specimen	NA	NA	NA	Leaves and stems fall in large numbers, very spiny
Specimen	NA	NA	NA	
Specimen	NA	NA	NA	Slow growing
Grouped together	NA	NA	NA	Numerous large cones to pick up
Specimen, shade	NA	NA	NA	Very slow to get started when transplanted

(continued)

and shrubs is particularly arbitrary. Some trees are used as shrubs by continued pruning, and *vice versa*. By and large the categories are all right, and thus they are useful in planning and landscaping.

3. Leaf Retention: **D** = deciduous; **E** = evergreen.
4. Light Preference: **S** = full sun; **SH** = full shade; **PS** = partial shade; **I** = indifferent, no apparent preference.
5. Major Horticultural Feature: this refers only to the ornamental use. Some of the included plants also have commercial or food value.
6. Flowering Time: **ES** = early spring; **S** = spring; **LS** = late spring; **SU** = summer; **LSU** = late summer; **F** = fall; **W** = winter; **U** = unknown; **NA** = not applicable, plants do not bear true flowers *e.g.* the pines.
7. Flower and fruit colors: **NA** = either flower or fruit is not important in the use of the plant for ornamental purposes or the plant does not produce flowers or fruit; **?** = colors unknown.

Tall Trees (90' or taller) (continued)

Plant Name[1]	Ht. at Maturity[2]	Leaf Retention[3]	Light Preference[4]	Major Horticultural Features[5]
Mockernut Hickory *Carya tomentosa*	100'	D	S-SH	Bright yellow fall foliage
Old Field Pine *Pinus taeda*	100'	E	S-PS	Massive trunks; short needles
Pecan *Carya illinoinensis*	150'	D	S	Tall, rounded crown
Pignut Hickory *Carya glabra*	100'	D	S-PS	Bright yellow fall foliage
Shortleaf Yellow Pine *Pinus echinata*	100'	E	S	Mostly for reforestation
Slash Pine *Pinus elliottii*	90'	E	S	Long, glistening needles
Spruce Pine *Pinus glabra*	100'	E	PS	Smooth bark; small cones
Sweet Gum *Liquidambar styraciflua*	100'	D	I	Handsome, star-shaped leaves
Sycamore *Platanus occidentalis*	150'	D	S	Striking, light-colored trunks
Tulip Tree *Liriodendron tulipifera*	130'	D	S-SH	Stately tree; attractive flowers and leaves
White Ash *Fraxinus americana*	120'	D	I	Slender, pyramidal shape
White Cedar *Chamaecyparis thyoides*	90'	E	PS-S	Trunk gray; crown pyramidal when young, round at maturity
White Pine *Pinus strobus*	120'	E	PS	Silver-green, flexible needles

Major Use	Flower Time[6]	Flower Color[7]	Fruit Color[7]	Undesirable Attributes
Background	S	?	Brown	Rather slow-growing tree
Grouped, light shade	NA	NA	NA	
Singly or groves	LS	Green	Brown	Huge production of pollen-bearing catkins
Specimen, shade	S	Green	Dark brown	Large, hard nuts, difficult and dangerous to mow over
Shade	NA	NA	NA	
Shade	NA	NA	NA	
Shade	NA	NA	NA	
Specimen, avenue tree	ES	Green	Brown	Numerous dry, sharp-pointed globular fruits hard to clean up
Shade, street tree	S	Green	Brown	Numerous large brown fruits; fallen leaves plaster the ground
Shade tree	S	Yellow-green	Brown	
Shade	S	Green	Tan	
Accent	NA	NA	NA	
Specimen	NA	NA	NA	

Medium Trees (45'–90')

Plant Name	Ht. at Maturity	Leaf Retention	Light Preference	Major Horticultural Features
American Holly *Ilex opaca*	50'	E	S-SH	Bright red clusters of berries at Christmas
Arbor Vitae *Thuja occidentalis*	60'	E	S-PS	Symmetrical shape; flat fan-like sprays of leaves
Black Locust *Robinia pseudoacacia*	80'	D	I	Very hardy; fragrant flowers
Black Willow *Salix nigra*	50'	D	SH-S	Attractive, rounded shape; yellow in fall
Cabbage Palm *Sabal palmetto*	60'	E	S	Large, graceful fan-shaped fronds
Chinaberry *Melia azedarach*	50'	D	S	Globe-shaped tree
Chinese Chestnut *Castanea mollissima*	60'	D	S	Attractive leaf shape
Chinese Evergreen Oak *Quercus myrsinifolia*	60'	E	S-PS	Conical tree shape
Chinese Parasol Tree *Firmiana simplex*	60'	D	S-PS	Large leaves; green bark
Devilwood *Osmanthus americanus*	45'	E	PS	Glossy foliage; fragrant flowers
East Palatka Holly *Ilex × attenuata*	45'	E	PS	Nearly spineless leaf; red berries
Eastern Hemlock *Tsuga caroliniana*	75'	E	S-PS	Narrow conical shape
Eastern Red Cedar *Juniperus virginiana*	50'	E	I	Dense foliage; may be pruned to fanciful shape

Major Use	Flower Time	Flower Color	Fruit Color	Undesirable Attributes
Display, food for birds	S	White	Red	Dead spiny leaves fall off in spring
Borders, basal plantings, pruned	NA	NA	NA	
Specimen or borders	S	White	Brown	Thorny branches
Accent	S	Yellow	?	Females produce bushels of hairy seeds; short lived
Specimen, avenue tree	SU	Green	?	Quite sensitive to cold
Accent, shade	S	Lavender	Yellow	Bitter tasting, poisonous (?) fruits
Shade	S	White	Brown	Flowers have disagreeable, sickening-sweet aroma
Avenue tree, accent	S	Tan	?	
Accent, shade	SU	Light brown	Brown	Spreads rapidly by runners and seedlings
Corners, accent	S	Creamy	Dark purple	
Accent	S	White	Red	
Specimen	NA	NA	NA	
Borders, specimen	NA	NA	NA	

(continued)

Medium Trees (45'–90') (continued)

Plant Name	Ht. at Maturity	Leaf Retention	Light Preference	Major Horticultural Features
Japanese Yew *Podocarpus macrophyllus*	45'	E	S-PS	Densely plated, strap-shaped leaves
Kwanzan Cherry *Prunus serrulata* cv. Kwanzan	60'	D	S	Profusion of flowers
Laurel Oak *Quercus laurifolia*	75'	D	I	Rapid growth
Live Oak *Quercus virginiana*	80'	E	S-PS	Beautiful stately tree with wide-spreading limbs
Loblolly Bay *Gordonia lasianthus*	70'	E	PS	Showy flowers, dark green foliage
Lombardy Poplar *Populus nigra* cv. Italica	60'	D	S	Columnar growth
Pond Cypress *Taxodium ascendens*	75'	D	PS	Tall, broad-based trunks
Poverty Pine *Pinus virginiana*	50'	E	S	Bushy, short needles
Princess Tree *Paulownia tomentosa*	60'	D	S	Large leaves, clusters of purple flowers
Red Maple *Acer rubrum*	50'	D	PS	Bright red flowers and fruit in early spring
River Birch *Betula nigra*	60'	D	PS	Handsome bark and foliage
Sand Pear *Pyrus pyrifolia*	50'	D	S	Decorative early spring flowers
Sand Pine *Pinus clausa*	70'	E	S	Bright green, bushy foliage

Major Use	Flower Time	Flower Color	Fruit Color	Undesirable Attributes
Foundation or accent, pruned	NA	NA	NA	
Specimen or avenue tree	S	White-lt. pink	?	
Shade	ES	Brown	Brown	Huge pollen and acorn production
Specimen	ES	Brown	Brown	
Display	SU	White	?	
Avenue tree, display	?	?	?	Dies back early (15–20 years)
Accent	NA	NA	NA	
Borders, pruned to hedge	NA	NA	NA	
Specimen	S	Purple	Brown	
Display	ES	Red	Red	
Accent	S	Green	?	
Background	ES	White	Yellow-brown	Fruits, though edible, are gritty
Borders, background	NA	NA	NA	

(continued)

Medium Trees (45'–90') (continued)

Plant Name	Ht. at Maturity	Leaf Retention	Light Preference	Major Horticultural Features
Sassafras *Sassafras albidum*	50'	D	PS-SH	Rapid growth, interesting leaf shape, red-yellow fall foliage
Silver Bay *Magnolia virginiana* var. *australis*	60'	E	PS-SH	Leaves silvery below; numerous fragrant, white flowers
Silver Dollar Tree *Eucalyptus polyanthemos*	50'	E	S	Rounded juvenile foliage
Sour Gum *Nyssa sylvatica*	55'	D	PS	Foliage glossy green; bright colors in fall
Sourwood *Oxydendron arboreum*	40'	D	PS	Leaves red, yellow in fall
Southern Magnolia *Magnolia grandiflora*	80'	E	S-SH	Stately tree; large, white, fragrant flowers
Southern Red Oak *Quercus falcata*	80'	D	S	Rounded crown, vigorous
Washington Palm *Washingtonia filifera*	50'	E	S-PS	Large, shiny, fan-shaped leaves
Water Oak *Quercus nigra*	80'	D	S-PS	Holds foliage until January
White Mulberry *Morus alba*	80'	D	PS-SH	Fast growth, attractive to birds
Worm Tree *Catalpa bignonioides*	50'	D	I	Large flower clusters

Major Use	Flower Time	Flower Color	Fruit Color	Undesirable Attributes
Accent	ES	Green	Red	Reproduces rapidly by runners, seeds
Accent	S-SU	White	Red	
Display	?	?	?	Cold sensitive
Specimen	LS	?	Dark blue	
Specimen	S	White	?	
Display, shade	S-SU	White	Brown	Constant leaf fall
Shade	S	Brown	Brown	
Accent, specimen, avenue tree	S	Light yellow	?	Large drooping "skirt" of dead leaves if not trimmed
Shade	ES	Brown	Brown	Huge pollen and acorn production
Background	S	?	White to purple	Fruit fall very messy
Background	S	White	Brown	

Small Trees (10'–45')

Plant Name	Ht. at Maturity	Leaf Retention	Light Preference	Major Horticultural Features
Apple *Malus sylvestris*	20'	D	S	Masses of flowers
Ashe's Magnolia *Magnolia ashei*	15'	D	PS	Large leaves silvery below; large fragrant flowers
Blackjack Oak *Quercus marilandica*	30'	D	S-PS	Handsome glossy foliage
Bradford Pear *Pyrus calleryana*	35'	D	S	Masses of flowers before leaves
Burford Holly *Ilex cornuta* cv. Burfordii	12'	E	S-PS	Covered with bright red berries in fall, winter
Cherry Laurel *Prunus caroliniana*	25'	E	PS	Glossy foliage
Chickasaw Plum *Prunus angustifolia*	16'	D	S	Numbers of red or yellow small fruits
Chinese Juniper *Juniperus chinensis*	40'	E	S	Attractive branching and foliage
Chinquapin *Castania pumila*	20'	D	S-PS	Glossy foliage, rounded crown
Columnar White Cedar *Chamaecyparis thyoides* cv. Ericoides	40'	E	S	Columnar form
Common Jujube *Ziziphus jujuba*	40'	D	I	Attractive olive-shaped fruit
Crabapple *Malus*, several sp.	12'–15'	D	S-PS	Bright flowers, red fruits

Major Use	Flower Time	Flower Color	Fruit Color	Undesirable Attributes
Accent, border	S	White	Various	Subject to fireblight, needs frequent spraying
Display	S	White	Red-brown	
Background, borders	S	Green	Brown	
Specimen, avenue tree	S	White	Brown	Flowers have rather fetid odor
Display, borders	SU	White	Bright red	
Background, pruned as hedge	S	Creamy	Purple	Produces large amount of fruits, birds scatter seed, many seedlings sprout up
Hedgerow	S	White	Red or yellow	
Accent	NA	NA	NA	
Background	S	White	Brown	
In rows, or accent	NA	NA	NA	Cold sensitive; turns brown in patches, leaving uneven color
Background tree	S	Greenish	Shiny reddish brown	Small thorns at each joint of branches
Display	S	Red, pink, white	Red	Regular fungal and insecticide sprays required for best results

(continued)

Small Trees (10'–45') (continued)

Plant Name	Ht. at Maturity	Leaf Retention	Light Preference	Major Horticultural Features
Crape Myrtle *Lagerstroemia indica*	18'	D	S	Full flower midsummer–early fall; red, pink, white, or lavender flowers
Dogwood *Cornus florida*	40'	D	PS	Great floral display, bright red berries
Fragrant Olive *Osmanthus fragrans*	30'	E	S	Fragrant small white flowers
Franklin Tree *Franklinia alatamaha*	30'	D	S-PS	Handsome white flowers
Glossy Privet *Ligustrum lucidum*	30'	E	S	Foliage, pruned
Grancy Graybeard *Chionanthus virginicus*	30'	D	PS	Cascades of fringy, white flowers with young, bright green leaves
Hawthorn *Crataegus uniflora*	10'	D	S	Attractive twisted trunk, many white flowers
Japanese Maple *Acer palmatum*	20'	D	S-PS	Deeply dissected red leaves
Japanese Persimmon *Diospyros kaki*	15'	D	S	Large glossy fruits
Jerusalem Thorn *Parkinsonia aculeata*	20'	D	S	Feathery foliage, numerous yellow flowers
Kumquat, Oval *Fortunella margarita*	12'	E	S	Glossy foliage, bright orange fruit
Loquat *Eriobotrya japonica*	25'	E	S	Tree has interesting shape; rough-surfaced leaves
Mimosa *Albizia julibrissin*	30'	D	S-PS	Gray bark; feathery foliage; powder-puff flowers

Major Use	Flower Time	Flower Color	Fruit Color	Undesirable Attributes
Display, borders	SU-EF	Red, lavender pink, white	?	
Accent, border, specimen tree	S	White, pink, red	Red	
Background, borders	ES	White	?	
Specimen	S-SU	White	?	
Hedge, accent	S	White	Dark blue to purple	Attracts white fly
Display	S	White	Dark purple	
Accent	S	White	?	Has scratchy thorns
Display	S	Red	Red	
Specimen	S	?	Reddish-yellow-pink	
Display	S	Yellow	Brown	Numbers of thorns along branches
Specimen	S	White	Orange	Somewhat cold sensitive, requires protection
Accent, corners	S	?	Orange	Flowers frequently killed by frost
Display	S-SU	Pink	Brown	Relatively short-lived tree

(*continued*)

Small Trees (10'–45') (continued)

Plant Name	Ht. at Maturity	Leaf Retention	Light Preference	Major Horticultural Features
Photinia *Photinia serrulata*	20'	E	S	Foliage and flowers
Pindo Palm *Butia capitata*	20'	E	S	Large gray-green fronds
Popcorn Tree *Sapium sebiferum*	40'	D	I	Fast-growing, red to yellow fall foliage
Pyramidal Magnolia *Magnolia pyramidata*	35'	D	PS-SH	Interesting whorled-leaf arrangement; large, white, fragrant flowers
Redbud *Cercis canadensis*	30'	D	PS-SH	Very early spring flowers
Red Tip (Top) *Photinia × fraseri*	25'	E	S-PS	Bright red young foliage, excellent pruned hedge
Sand Live Oak *Quercus virginiana* var. *maritima*	25'	E	S	Twisted branching, evergreen leaves
Satsuma *Citrus reticulata*	15'	E	S	Dark green foliage, orange fruit
Saucer Magnolia *Magnolia × soulangeana*	25'	D	PS	Early flowering before leaves
Silver-bell Tree *Halesia diptera*	30'	D	PS	Numerous pendant flowers on every branch
Sloe *Prunus americana* or *alleghaniensis*	25'	D	PS	Early flowering, fragrant
Smoke Tree *Cotinus coggygria*	15'	D	S-PS	Inflorescenses give appearance of "puff of smoke"

Major Use	Flower Time	Flower Color	Fruit Color	Undesirable Attributes
Borders	S	White	?	
Display	SU	Whitish	Orange-yellow	
Accent	S	Yellow-green	White	Sends out numerous runners
Display	S	White	Reddish	
Accent, display	ES	Pink	Brown	
Hedges	S	White	?	
Accent	?	?	?	
Specimen	ES	White	Orange	Cold sensitive, attracts white fly
Display	ES	White-pink to reddish	?	
Display	ES	White	Brown	
Accent	ES	White	?	
Display	S	Pinkish	?	

(*continued*)

Small Trees (10'–45') *(continued)*

Plant Name	Ht. at Maturity	Leaf Retention	Light Preference	Major Horticultural Features
Snowdrop Tree *Halesia carolina*	40'	D	PS	Early, pendant flowers
Star Magnolia *Magnolia stellata*	20'	D	PS	Early white flowers
Taiwan Cherry *Prunus campanulata*	30'	D	S	Earliest pink flowers
Ternstroemia *Ternstroemia gymnanthera*	20'	E	PS	Glossy leaves, flowers yellow-white
Tung Oil Tree *Aleurites fordii*	20'	D	S	Numerous large flowers, rounded crowns
Weeping Willow *Salix babylonica*	30'	D	S	Graceful drooping branches, fast growing
Wild Cassava *Manihot grahamii*	15'	D	S	Attractive, rounded crown and leaf form
Wild Persimmon *Diospyros virginiana*	25'	D	S	Deep red or yellow fall color, orange fruit
Windmill Palm *Trachycarpus fortunei*	40'	E	S-PS	Good palm for small gardens

Major Use	Flower Time	Flower Color	Fruit Color	Undesirable Attributes
Display	ES	White	Brown	
Accent	ES	White	?	
Display	ES	Pink	Brown	
Border, accents	S	Yellow-white	Brown with bright red seed	
Specimen	S	White-pink	Brown	Large fruit with poisonous seeds
Display	ES	?	?	
Display	LS	Yellow-green	Brown	Cold sensitive
Borders	S	?	Orange	
Accent	?	?	?	

Shrubs

Plant Name	Ht. at Maturity	Leaf Retention	Light Preference	Major Horticultural Features
Abelia *Abelia × grandiflora*	5–6'	E	I	Continuous summer flowering
Alabama Azalea *Rhododendron alabamense*	3–4'	E	PS	Pure white, scented flowers
Althea *Hibiscus syriacus*	15'	D	S	Vigorous growth, variety of flower colors
Aucuba *Aucuba japonica*	15'	E	SH	Variegated or plain green foliage
Bamboo (Fishpole) *Phyllostachys aurea*	25'	E	S-PS	Thick-growing clusters
Banana Shrub *Michelia figo*	15'	E	S-PS	Pleasing globular shape, numerous yellow flowers
Beargrass *Yucca flaccida*	1.5'	E	S	3–4 foot spikes of white flowers
Beautyberry *Callicarpa americana*	6'	D	I	Sequential clusters of bright purple berries along stems
Boxwood (Japanese) *Buxcus microphylla* var. *japonica*	3–4'	E	S	Elegant hedge plants
Burning Bush *Euonymus atropurpureus*	20'	D	PS	Good fall color
Butterfly Bush *Buddleja davidii*	15'	D	S	Long clusters of bright blue flowers
Camellia (Japanese) *Camellia japonica*	15–20'	E	PS	Tremendous array of floral form and color
Camellia (Sasanqua) *Camellia sasanqua*	12'	E	PS	Numerous white to pink flowers

Major Use	Flower Time	Flower Color	Fruit Color	Undesirable Attributes
Singly, in groups, hedges	S-F	White-pink	?	
Accent	ES	White	?	
Accents, borders	SU	White, pink, red	?	
Singly	?	?	Red	
Hedges	NA	NA	NA	Difficult to root out; spreads rapidly by runner
Singly	LS	Yellow	?	
Groupings, borders	LS-SU	White	Brown	
Singly	LS	White	Reddish purple	
Hedges	?	?	?	Slow growing, susceptible to nematodes in sandy soils
Singly or hedges	S	Purple	Crimson	
Display	SU	Blue	?	Somewhat tender
Specimen, borders	F-W-ES	White to red, variegated	Brown	
Display	LSU-F	White or pink	Brown	

(continued)

Shrubs (continued)

Plant Name	Ht. at Maturity	Leaf Retention	Light Preference	Major Horticultural Features
Cassia *Cassia corymbosa*	8'	E	S	Bright flowers, late summer and fall
Cassia *Cassia coluteoides*	15'	E	S	Wider leaves than *C. corymbosa*
Century Plant *Agave americana*	6'	E	S	Large fleshy leaves in basal rosette
Chapman's Azalea *Rhododendron chapmanii*	6'	E	PS	Trumpet-shaped flowers
Chinese Hibiscus *Hibiscus rosa-sinensis*	8'	E	S	Large single or double flowers
Common privet *Ligustrum sinense*	15'	E	I	Vigorous grower
Confederate Rose *Hibiscus mutabilis*	15'	D	S-PS	Each double flower changes from white to pink in one to two hours
Coontie *Zamia pumila*	2–3'	E	PS	Stiff, leathery fronds
Crape jasmine *Tabernaemontana divaricata*	8'	E	S-PS	Glossy leaves, white fragrant flowers
Crimson Bottlebrush *Callistemon citrinus*	20'	E	S	Inflorescences look like 8 inch bottlebrushes
Deutzia *Deutzia scabra* cv. Candidissima	15'	D	S-PS	Bountiful white flowers in spring
Dwarf Gardenia *Gardenia jasminoides* cv. Radicens	2'	E	S	Covered with white, fragrant blooms

Major Use	Flower Time	Flower Color	Fruit Color	Undesirable Attributes
Accent	LSU-F	Yellow	Brown	
Accent	LSU	Yellow	Brown	A bit more sensitive to frost than *C. corymbosa*
Display	?	Yellow or reddish	Brown	Flowers only once in variable lifetime, then dies
Accent	LS	Rose-pink	?	
Display, borders	SU	White to red-yellow	?	Must be protected in winter
Hedges	S-SU	White	Blue-black	Harbors clouds of white fly; spreads rapidly
Display	LSU F	White to pink	Brown	
Singly or borders	NA	NA	NA	
Accent	SU	White	?	Frost-sensitive, dies back in winter
Singly, borders	S	Red	?	Seldom allowed to get much more than 8 feet tall
Display	S	White	?	Tendency to become leggy unless pruned every 2–3 years
Borders	S	White	?	Must be sprayed for white fly

(continued)

Shrubs (continued)

Plant Name	Ht. at Maturity	Leaf Retention	Light Preference	Major Horticultural Features
European Fan Palm *Chamaerops humilis*	8'	E	S	Low-growing palm
Fatsia *Fatsia japonica*	20'	E	SH	Attractive large glossy leaves
Feijoa *Feijoa sellowiana*	18'	E	S	Good foliage, interesting flowers
Fetterbush *Leucothoe populifolia*	12'	E	PS	Erect shrubs with white, blueberry-like flowers
Firethorn (scarlet) *Pyracantha coccinea*	8'	E	S	Bright red berried shrub either as bush or espaliered
Firethorn (crenulate) *Pyracantha crenulata*	20'	E	S	Orange-red berries
Florida Anise *Illicium floridanum*	10'	E	PS-SH	Shade-loving, red flowers
Florida Flame Azalea *Rhododendron austrinum*	15'	D	PS	Early, yellow-orange flowers
Florida Yew *Taxus floridana*	20'	E	S-PS	Pruned shrub
Flowering Almond *Prunus glandulosa* cv. Sinensis	3'	D	S-PS	Early-flowering pink
Flowering Quince *Chaenomeles speciosa*	6'	D	S	Flowers early before leaves
Gardenia *Gardenia jasminoides*	6'	E	PS	Double white flowers, delightful fragrance
Giant Reed *Arundo donax*	18'	E	S	Long stems with green or variegated leaves

Major Use	Flower Time	Flower Color	Fruit Color	Undesirable Attributes
Singly	?	?	?	
Corners	?	White	?	
Hedges, borders, single	S	White, red stamens	Green, some red	
Corners	S	White	?	
Free-standing or espallier	S	White	Red	Susceptible to fire blight
Free-standing or espallier	S	White	Orange-red	
Display	S	Red	Brown	
Display, borders	ES	Yellow-orange	Brown	
Background	NA	NA	NA	Difficult to establish
Borders	ES	Pink	?	
Specimen	ES	Orange-red	Brown-red	
Specimen	S	White	?	
Entrance-ways, corners	S	Tan	?	

257

(continued)

Shrubs (continued)

Plant Name	Ht. at Maturity	Leaf Retention	Light Preference	Major Horticultural Features
Golden Bells *Forsythia × intermedia*	8'	D	S-PS	Arching branches with yellow flowers
Highbush Huckleberry *Vaccinium corymbosum*	16'	D	SH	Light green foliage early in the spring
Honeysuckle Bush *Rhododendron canescens*	10'	D	PS	Flower before leaves
Hydrangea (Hortensia) *Hydrangea macrophylla* var. *macrophylla*	4'	D	S-PS	Large rounded trusses of blue (acid) or pink (basic) sterile flowers
Hydrangea (Lace Cap) *Hydrangea macrophylla* var. *macrophylla*	4'	D	S-PS	Fringe of sterile white or blue flowers around inflorescence
Indian Hawthorn *Rhaphiolepis indica*	8'	E	S	Rosy-pink flowers, leathery leaves
Japanese (Chinese) Jasmine *Jasminum mesnyi*	10'	E	S	Makes large rounded clump with yellow flowers
Japanese Pittosporum *Pittosporum tobira*	18'	E	S	Pruned plants with glossy foliage
Kumquat (Round) *Fortunella japonica*	8'	E	S	Glossy foliage, orange fruits
Lantana *Lantana camara*	4'	D	S	Rank-growing
Ligustrum *Ligustrum japonicum*	10'	E	S	Vigorous grower, nice leaf pattern (variegated)
Lily-flowered Magnolia *Magnolia liliiflora*	8'	D	PS	Very early flowering before leaves; striking flowers

Major Use	Flower Time	Flower Color	Fruit Color	Undesirable Attributes
Accent, borders	S	Yellow	?	
Background	ES	White	Blue	
Accent	ES	White with pink	?	
Borders, specimens	SU	Blue or pink		
Borders	SU	Blue or white	?	
Borders, accent	S-SU	Rose-pink	?	
Specimen	S	Yellow	?	
Borders	S	White	Purple	
Accent, specimen	S	White	Orange	Rather tender, needs covering in coldest weather
?	S	Yellow, red, white	?	Unpleasant odor, weedy tendency
Hedges, background	SU	White	?	
Specimen	ES	Purple, white inside	Brown with bright red seeds	

(continued)

Shrubs (continued)

Plant Name	Ht. at Maturity	Leaf Retention	Light Preference	Major Horticultural Features
Mallow Rose *Hibiscus moscheutus*	6'	D	S	Large, bright flowers in profusion
Mandarin Orange *Citrus reticulata*	10'	E	S	Glossy foliage; fragrant, white flowers; orange fruit
Mock Orange *Philadelphus coronarius*	6'	D	PS	Nice, white, fragrant flowers
Mountain Laurel *Kalmia latifolia*	15'	E	PS	Early-flowering shrubs
Nandina *Nandina domestica*	7'	E	I	Long-lasting, large clusters of red berries
Needle Palm *Rhapidophyllum hystrix*	6'	E	SH	Glossy fronds, low-growing
Oakleaf Hydrangea *Hydrangea quercifolia*	15'	D	S-PS	Large, white flower trusses
Oleander *Nerium oleander*	10'	E	S	Handsome foliage and flowers
Oregon Grape (Leatherleaf) *Mahonia bealei*	3'	E	PS	Holly-like leaves, clusters of yellow flowers
Plumleaf Azalea *Rhododendron prunifolium*	10'	D	PS	Summer-flowering
Pomegranate *Punica granatum*	20'	D	S	Bright, orange-red flowers
Prickly Pear *Opuntia ficus-indica*	4	E	S	Cactus plant with large, oval pads
Red Buckeye *Aesculus pavia*	12'	D	SH	Early, bright red flowers

Major Use	Flower Time	Flower Color	Fruit Color	Undesirable Attributes
Specimen	SU	White, pink, or red	Brown	
Specimen, fruit	SU	White	Orange	
Accent	S	White	?	
Specimen	ES	Pink	?	
Borders, accent	SU	White	Red	
Specimen	S	?	?	
Accent	SU	White	?	Tendency to spread by runners
Borders	SU	Red, white, or pink	?	Foliage poisonous, fresh or burned
Accent	F?	Yellow	Blue-black	
Specimen	LSU	Orange-red	?	
Accent	S	Orange-red	Reddish	
Corners	SU	Yellow-green	Purple-red; green	Very spiny; don't plant near foot traffic
Accent	ES	Red	Brown	

(continued)

Shrubs (continued)

Plant Name	Ht. at Maturity	Leaf Retention	Light Preference	Major Horticultural Features
Rice Paper Plant *Tetrapanax papyriferus*	12'	E	PS	Large foliage
Sago Palm *Cycas revoluta*	10'	E	I	Dark green, glossy fronds
Scarlet Wisteria Tree *Sesbania punicea*	8'	D	S-PS	Large clusters of bright, orange-red flowers
Silky Camellia *Stewartia malacodendron*	15'	D	S	Large, frilly white flowers with large numbers of purple stamens
Silver Thorn *Elaeagnus pungens* cv. Simonii	15'	E	S	Vigorous, thick growth; gray-green foliage; thorns on stem
Snowbell *Styrax americanus*	10'	D	PS	Numerous, pendulant white flowers
Spanish Bayonet *Yucca aloifolia*	12'	E	S	Handsome white flowers, thorny leaves
Sparkleberry *Vaccinium arboreum*	15'	E	SH	Shiny leaves and black berries, white flowers; fruits attract birds
Spindle Tree *Euonymus japonicus*	8'	E	S-PS	Handsome foliage, green or variegated
Spiraea *Spiraea × bumalda*	4'	D	S	Flat-topped flower clusters
Spiraea *Spiraea cantoniensis*	8'	D	S	Early, double, white flowers
Spiraea *Spiraea prunifolia*	7'	D	S	Early double flowers
Spiraea *Spiraea thunbergii*	8'	D	S	Clusters of very small, white flowers

Major Use	Flower Time	Flower Color	Fruit Color	Undesirable Attributes
Corners	W	White	?	
Corners, borders	NA	NA	NA	
Accent	S	Orange-red	Brown, winged	
Specimen	S	White	?	
Hedges	SU	Tan	Brown	
Borders	S	White	Brown	
Hedges, fences	S-SU	White	Green, brown	
Background, native border	S	White	Blue-black	
Basal plantings, accent	?	?	?	
Borders	SU	Pink or white	?	
Accent	ES	White	?	
Borders	ES	White	?	
Specimen	ES	White	?	

(*continued*)

Shrubs (continued)

Plant Name	Ht. at Maturity	Leaf Retention	Light Preference	Major Horticultural Features
Strawberry Bush *Euonymus americanus*	8'	E	PS	Attractive fruit
Sweet Shrub *Calycanthus floridus*	6'	D	PS	Sweet, dark red flowers, fruits balloon-shaped
Texas Sage *Leucophyllum frutescens*	8'	E	S	Grey-green foliage, rosy pink flowers
Turk's Cap Hibiscus *Malvaviscus arboreus* var. *mexicanus*	4'	D	S	Late-blooming flowers
Wax Myrtle *Myrica cerifera*	15'	E	PS-SH	Rounded form, pungent evergreen leaves
Weigela *Weigela florida*	10'	D	S	Profuse flowering
Wintersweet *Chimonanthus praecox*	10'	D	PS	Welcome winter flowers, fragrant
Yaupon *Ilex vomitoria*	25'	E	I	Bright red berries at Christmas time

Major Use	Flower Time	Flower Color	Fruit Color	Undesirable Attributes
Accent	S	Greenish	Bright red	
Borders	ES	Red	Brown	Spreads rapidly by runners
Accent	SU	Rosy-pink	?	
Specimen	EF	Red	?	
Borders, hedges	?	?	Bluish bloom	
Accent	S	Reddish purple	?	
Specimen	W	Yellow outside, striped purple inside	?	
Hedges, borders	S-SU	White	Bright red berries	

Woody Vines

Plant Name	Leaf Retention	Plant Part of Major Horticultural Value	Description of Plant
Algerian Ivy *Hedera canariensis*	E	Leaves	Leaves bright green with red petioles
Argentine Trumpet Vine *Clytostoma callistegioides*	E	Tubular flowers	Leaves elliptic, 4 inches long; flowers light purple with white throat, flowering in early summer
Confederate Jasmine *Trachelospermum jasminoides*	E	Fragrant white flowers	Leaves opposite, elliptic, 2 inches long, glossy, leathery; plants climb by twining
English Ivy *Hedera helix*	E	Juvenile foliage most decorative	Shiny leaves with green petioles differentiates this one from Algerian ivy; climbs by modified tendrils
Honeysuckle (vine) *Lonicera japonica*	E	Flowers and foliage	Plants climb by twining; very aggressive grower, also spread by birds eating fruits; very nice fragrance but a weed in most places
Muscadine Grape *Vitis rotundifolia*	D	Fruits are edible and decorative, bronze and purple varieties	Climbs by tendrils; predominant grape in our area; grapes separate easily from the clusters
Trumpet Creeper *Campsis radicans*	D	Tubular, orange-red flowers, 2 inches across, 3 inches long	Climbs by aerial rootlets; leaves compound, each with 5–9 leaflets; flowers from midsummer till fall; attractive to hummingbirds; leaves and flowers said to be poisonous
Trumpet Honeysuckle *Lonicera sempervirens*	E	Bright red flowers, related to other honeysuckle	Similar to its relative above in foliage, but does not spread and become weedy.

Plant Name	Leaf Retention	Plant Part of Major Horticultural Value	Description of Plant
Virginia Creeper *Parthenocissus quinquefolia*	D	Palmately compound leaves turning red in fall	Climbs by modified tendrils; vigorous grower, with small inconspicuous flowers
Wisteria *Wisteria floribunda* and/or *W. sinensis*	D	Large clusters of white or purple fragrant flowers	Climbs by twining; very vigorous plants, escaped from cultivation; plants reproduce by runners and by seeds; two species are practically interchangeable
Yellow Jasmine *Gelsemium sempervirens*	E	Individual, yellow, fragrant flowers distributed along stems	Climbs by twining; long flowering period; foliage is poisonous

Leaf retention: **D** = deciduous; **E** = evergreen

Woody Ground Covers

Plant Name	Leaf Retention	Description of Plant
Common Juniper *Juniperus communis*	Evergreen	The prostrate forms of this *Juniperus* species are found under many different names, but all are useful in one sunny place or another: in borders in front of perennials, along sidewalks, in circular beds with a central, taller juniper, etc.
English Ivy *Hedera helix*	Evergreen	Though listed previously under vines, this ivy is popularly used as a ground cover. Its vigorous growth, bright green foliage, and ground-hugging qualities are appreciated. However, if the plants reach any erect structure, they will begin to grow upward.
Shore Juniper *Juniperus conferta*	Evergreen	Vigorous ground cover for sunny, dry locations. Plants grow horizontally with upward-growing side branches reaching about one foot tall, but sometimes a little more. Foliage is of needle-form, each leaf about ½ inch long, closely set on stems, usually light green.
Winter Creeper *Euonymus fortunei*	Evergreen	This *Euonymus* species grows well in shade, in contrast to the two other ground covers listed above. There are many variations in leaf size and in colors, some variegated, others plain green. Not too commonly used.

Selected Reading List

Native Plants in the Garden

Clewell, A. F. 1988. *Guide to the Vascular Plants of the Florida Panhandle.* Tallahassee: Florida State University Press.

Correll, D. S., and M. C. Johnston. 1970. *Manual of the Vascular Plants of Texas.* Renner, Tex.: Texas Research Foundation.

Davis, Fanny-Fern. 1988. *Nature's Seasonal Splendor, Native Plants and Wildflowers, District One.* Winter Park: Florida Federation of Garden Clubs, Inc.

Duncan, W. H., and M. B. Duncan. 1988. *Trees of the South-eastern United States.* Athens: University of Georgia Press.

Godfrey, R. K. 1988. *Trees, Shrubs, and Woody Vines of Northern Florida and Adjacent Georgia and Alabama.* Athens: University of Georgia Press.

Kurz, H., and Godfrey R. K. 1962. *Trees of Northern Florida.* Gainesville: University of Florida Press.

Ward, D. B., ed. 1979. *Rare and Endangered Biota of Florida.* Vol. 5, *Plants.* Gainesville: University of Florida Press.

West, E., and Arnold, L. E. 1956. *The Native Trees of Florida.* Rev. ed. Gainesville: University of Florida Press.

Cultivated Plants

Bailey, L. H. 1944. *Manual of Cultivated Plants.* New York: Macmillan.

Bailey, L. H., and Ethel Zoe Bailey, initial compilers. 1976. *Hortus III. A*

Concise Dictionary of Plants Cultivated in the United States and Canada. Revised and expanded by the Staff of the Liberty Hyde Bailey Horotorium. New York: Macmillan.

Rehder, A. 1947. *Manual of Cultivated Trees and Shrubs.* New York: Macmillan.

Garden Guides

Garden and Landscape Staff, Southern Living Magazine. 1981. *Southern Living Garden Guide.* Birmingham, Ala.: Oxmoor House.

River Oaks Garden Club. 1975. *A Garden Book for Houston and the Gulf Coast.* Houston: Pacesetter Press.

Trustees Garden Club. 1988. *Garden Guide to the Lower South.* Savannah: Trustees Garden Club. Available from the Trustees' Garden Club, 2711 Abercorn Street, Savannah, Georgia 31405.

Watkins, J. V., and T. J. Sheehan. 1975. *Florida Landscape Plants.* Rev. ed. Gainesville: University of Florida Press.

Watkins, J. V., and H. S. Wolfe. 1986. *Your Florida Garden.* Fifth ed., Abridged. Gainesville: The University of Florida Press.

Other Information Resources

County Agricultural Extension Service. The home gardener can use the services of this agency in several ways. One may have soil tests made with assurance of accuracy at a minimal cost. In addition, one may ask the county agent either to identify a plant or to have it done by someone in the state agricultural system. The office of the Extension Service is usually well provided with brochures and pamphlets on a variety of subjects of interest to the home owner, such as soils, fertilizers, diseases, insect pests, and it usually has a number of publications on various ornamental plants. There are also sources of information on weeds and how to eradicate them.

The County Forestry Service. This agency, which is a part of the overall state forestry organization, will have valuable data on native trees and where to buy seedlings of certain species. The forestry service may also have printed information of value, either at the local or state level. A visit to the forestry office will usually provide considerable information and leads on where to get more.

Libraries and bookstores. There may be a separate gardening section in libraries and bookstores. Librarians are very helpful when asked to find out something or other about plants—"what, why, where, and how to" information.

Index of Common Names

Illustrations not on same page as text are designated by italic page numbers.

Abelia, 3, 252
  Abelia × *grandiflora*
Alabama Azalea
  see Azalea, Alabama
Algerian Ivy
  see Ivy, Algerian
Almond, Flowering, 169, 256
  Prunus glandulosa cv. Sinensis
Althea, 86, 252
  Hibiscus syriacus
American Holly
  see Holly, American
American Ivy
  see Ivy, American
American Mulberry
  see Beautyberry
American Olive
  see Devilwood
Anise, Florida, 96, 256
  Illicium floridanum
Apple, 124, *125*, 244
  Malus sylvestris

Arbor Vitae, 213, *214*, 238
  Thuja occidentalis
  (*see also* Cedar, White)
Argentine Trumpet Vine
  see Trumpet Vine, Argentine
Ash, White, 73, 236
  Fraxinus americana
Ashe's Magnolia
  see Magnolia, Ashe's
Atlantic White Cedar
  see Cedar, White
Aucuba, 13, *14*, 252
  Aucuba japonica
Australian Laurel
  see Pittosporum, Japanese
Azalea, Alabama, 186, *187*, 252
  Rhododendron alabamense
Azalea, Chapman's, 188, 254
  Rhododendron chapmanii
Azalea, Florida Flame, 187, *188*, 256
  Rhododendron austrinum

Azalea, Plum-leaved, 189, 260
  Rhododendron prunifolium
Bald Cypress
  see Cypress, Bald
Bamboo, Fishpole, 149, *150*, 252
  Phyllostachys aurea
Bamboo, Heavenly
  see Nandina
Banana Shrub, 130, *131*, 252
  Michelia figo
Barometer Bush
  see Sage, Texas
Bay, Bull
  see Magnolia, Southern
Bay, Loblolly, 78, *79*, 240
  Gordonia lasianthus
Bay, Silver (Sweet), 118, *119*, 136, 242
  Magnolia virginiana var. *australis*
Bayonet, Spanish, 227, *228*, 264
  Yucca aloifolia
Beargrass, 227, *229*, 252
  Yucca flaccida
Beautyberry, 18, *19*, 252
  Callicarpa americana
Bells, Golden, 68, *69*, 258
  Forsythia × *intermedia*
Birch, Black
  see Birch, River
Birch, Red
  see Birch, River
Birch, River, 14, *15*, 240
  Betula nigra
Black Birch,
  see Birch, River
Black Laurel
  see Laurel, Black
Black Locust
  see Locust, Black

Black Willow
  see Willow, Black
Blackjack Oak
  see Oak, Blackjack
Blue Palmetto
  see Palm, Needle
Blueberry, Highbush
  see Huckleberry, Highbush
Blueberry, Swamp
  see Huckleberry, Highbush
Bottle Tree, Chinese
  see Parasol Tree, Chinese
Bottlebrush, Crimson, 19, *20*, 254
  Callistemon citrinus or *C. viminalis*
Boxwood, Japanese, 17, 252
  Buxus microphylla var. *japonica*
Bradford Pear
  see Pear, Bradford
Bridal Wreath (used for all the following species)
  Spiraea cantoniensis, S. prunifolia, S. thunbergii, S. × *bumalda*
Buckeye, Red, 6, *7*, 260
  Aesculus pavia
Bull Bay
  see Magnolia, Southern
Burford Holly
  see Holly, Burford
Burning Bush, 62, 262
  Euonymus atropurpureus
Bursting Heart
  see Strawberry Bush
Bush Honeysuckle
  see Honeysuckle (bush)
Butterfly Bush, 16, 252
  Buddleja davidii
Cabbage Palm
  see Palm, Cabbage
Calico Bush
  see Laurel, Mountain

Camellia, Japanese, 22, 252
  Camellia japonica
Camellia, Sasanqua, 23, 252
  Camellia sasanqua
Camellia, Silky, 204, *206*, 262
  Stewartia malacodendron
Camphor Tree, 44, 234
  Cinnamomum camphora
Canary Ivy
  see Ivy, Algerian
Cape Jasmine
  see Gardenia, Common
Carolina Jasmine
  see Jasmine, Yellow
Cassava, Wild, 127, *128*, 250
  Manihot grahamii
Cassia, 28, 254
  Cassia corymbosa or *C. coluteoides*, *29*, *30*
Catalpa, 33, *34*, 242
  Catalpa bignonioides
Cedar, Atlantic White
  see Cedar, White
Cedar, Eastern Red, 38, 99, *101*, 238
  Juniperus virginiana
Cedar, Japanese
  see Cryptomeria
Cedar, White, 38, *40*, 244
  Chamaecyparis thyoides
  see also Arbor Vitae
Ceniza
  see Sage, Texas
Century Plant, *8*, 254
  Agave americana
Chapman's Azalea
  see Azalea, Chapman's
Cherry, Japanese
  see Cherry, Kwanzan

Cherry, Kwanzan, 170, 240
  Prunus serrulata cv. Kwanzan
Cherry, Oriental
  see Cherry, Kwanzan
Cherry, Taiwan, 167, *168*, 250
  Prunus campanulata
Cherry Laurel
  see Laurel, Cherry
Chestnut, Chinese, 30, *31*, 238
  Castanea mollissima
Chickasaw Plum
  see Plum, Chickasaw
China Fir
  see Fir, China
China Rose
  see Hibiscus, Chinese
Chinaberry, 129, 238
  Melia azedarach
Chinese Bottle Tree
  see Parasol Tree, Chinese
Chinese Chestnut
  see Chestnut, Chinese
Chinese Date
  see Jujube, Common
Chinese Evergreen Oak
  see Oak, Chinese Evergreen
Chinese Hibiscus
  see Hibiscus, Chinese
Chinese Holly
  see Holly, Chinese
Chinese Jasmine
  see Jasmine, Chinese
Chinese Jujube
  see Jujube, Chinese
Chinese Juniper
  see Juniper, Chinese
Chinese Parasol Tree
  see Parasol Tree, Chinese
Chinese Privet
  see Privet, Glossy

Chinese Tallow Tree
  see Popcorn Tree
Chinquapin, 32, 244
  Castanea pumila
Cleyera
  see Ternstroemia
Common Gardenia
  see Gardenia, Common
Common Jujube
  see Jujube, Common
Common Juniper
  see Juniper, Common
Common Privet
  see Privet, Common
Common Rose Mallow
  see Rose Mallow, Common
Confederate Jasmine
  see Jasmine, Confederate
Confederate Rose
  see Rose, Confederate
Coontie, 230, 254
  Zamia pumila
Coral Honeysuckle
  see Honeysuckle, Trumpet
Crabapple, 124, *126*, 244
  Malus, several species
Crape Gardenia
  see Jasmine, Crape
Crape Jasmine
  see Jasmine, Crape
Crape Myrtle
  see Myrtle, Crape
Creeper, Trumpet, 24, 46, 266
  Campsis radicans
Creeper, Virginia, 143, *144*, 267
  Parthenocissus quinquefolia
Creeper, Winter, 62, 268
  Euonymus fortunei
Crimson Bottlebrush
  see Bottlebrush, Crimson

Cryptomeria, 51, *52*, 234
  Cryptomeria japonica
Cucumber Tree
  see Magnolia, Ashe's
Cypress, Bald, 209, 234
  Taxodium distichum
Cypress, Pond, 208, 240
  Taxodium ascendens
Date, Chinese
  see Jujube, Common
Deodar, 34, *35*, 234
  Cedrus deodara
Deutzia, 55, 254
  Deutzia scabra cv. Candidissima
Devil's Darning Needle
  see Beargrass
Devilwood, 138, *139*, 238
  Osmanthus americanus
Dogwood, 47, *48*, 246
  Cornus florida
Dogwood, English, 146, *147*
  Philadelphus inodorus
Dogwood, Flowering
  see Dogwood
Dwarf Gardenia
  see Gardenia, Dwarf
East Palatka Holly
  see Holly, East Palatka
Eastern Hemlock
  see Hemlock, Eastern
Eastern Red Cedar
  see Cedar, Eastern Red
Eastern Sycamore
  see Sycamore, Eastern
Eastern White Pine
  see Pine, White
Elaeagnus, Thorny
  see Silverthorn
English Dogwood
  see Dogwood, English

English Ivy
  see Ivy, English
European Fan Palm
  see Palm, European Fan
Evening Trumpet Flower
  see Jasmine, Yellow
Fatsia, Japanese, 64, *65*, 256
  Fatsia japonica
Feijoa, 64, *66*, 256
  Feijoa sellowiana
Fetterbush, 106, 256
  Leucothoe populifolia
Fir, China, 53, 234
  Cunninghamia lanceolata
Firethorn, 171, 256
  Pyracantha coccinea or *P. crenulata*, *172*, *173*
Fishpole Bamboo
  see Bamboo, Fishpole
Florida Anise
  see Anise, Florida
Florida Flame Azalea
  see Azalea, Florida Flame
Florida Yew
  see Yew, Florida
Flowering Almond
  see Almond, Flowering
Flowering Dogwood
  see Dogwood
Flowering Peach
  see Peach, Flowering
Flowering Quince
  see Quince, Flowering
Formosa Rice Paper
  see Fatsia, Japanese
Fragrant Olive
  see Olive, Fragrant
Franklin Tree, 71, 246
  Franklinia alatamaha
French Mulberry
  see Beautyberry

Fringe Tree
  see Graybeard, Grandsir (Grancy)
Gardenia, Common, 74, *75*, 256
  Gardenia jasminoides
Gardenia, Crape
  see Jasmine, Cape
Gardenia, Dwarf, 75, 254
  Gardenia jasminoides cv. Radicans
Giant Reed
  see Reed, Giant
Glossy Privet
  see Privet, Chinese
Gold Dust Shrub
  see Aucuba
Golden Bells
  see Bells, Golden
Grandsir (Grancy) Graybeard
  see Graybeard, Grandsir (Grancy)
Grape, Muscadine, 221, *222*, 266
  Vitis rotundifolia
Grape, Oregon
  see Leatherleaf
Graybeard, Grandsir (Grancy), 42, *43*, 246
  Chionanthus virginicus
Guava, Pineapple
  see Feijoa
Gum, Sour
  see Sour Gum
Haw, 50, *51*, 246
  Crataegus uniflora
Hawthorn
  see Haw
Hawthorn, Indian, 182, *183*, 258
  Rhaphiolepis indica
Heavenly Bamboo
  see Nandina
Hedge Plant
  see Privet, Common
Hemlock, Eastern, 217, *218*, 238
  Tsuga caroliniana

Hibiscus, Chinese, 85, 254
  Hibiscus rosa-sinensis
Hibiscus, Turk's Cap, 126, *128*, 264
  Malvaviscus arboreus var. *mexicanus*
Hickory, Mockernut, 27, 235
  Carya tomentosa
Hickory, Pignut, 25, *26*, 236
  Carya glabra
Highbush Blueberry (or Huckleberry)
  see Huckleberry, Highbush
Holly, American, 93, *94*, 238
  Ilex opaca
Holly, Burford, 92, *93*, 244
  Ilex cornuta cv. Burfordii
Holly, Chinese, 92
  Ilex cornuta
Holly, East Palatka, 91, 238
  Ilex × attenuata cv. East Palatka
Holly, Rotunda, 92
  Ilex cornuta cv. Rotunda
Holly, Mahonia
  see Leatherleaf
Honeysuckle (bush), 188, 258
  Rhododendron canescens
Honeysuckle (vine), 113, 266
  Lonicera japonica
Honeysuckle, Coral
  see Honeysuckle, Trumpet
Honeysuckle, Japanese
  see Honeysuckle (vine)
Honeysuckle, Trumpet, 114, 266
  Lonicera sempervirens
Hortensia Hydrangea
  see Hydrangea, Hortensia
Huckleberry, Highbush, 219, *220*, 258
  Vaccinium corymbosum
Hydrangea, Hortensia, 87, *88*, 258

Hydrangea macrophylla var. *macrophylla* cv. Hortensia
Hydrangea, Lace Cap, 88, *89*, 258
  Hydrangea macrophylla var. *macrophylla* cv. Lace Cap
Hydrangea, Oakleaf, 89, *90*, 260
  Hydrangea quercifolia
Indian Fig Cactus
  see Prickly Pear
Indian Hawthorn
  see Hawthorn, Indian
Ivy, Algerian, 81, 266
  Hedera canariensis
Ivy, American, 143, *144*, 267
  Parthenocissus quinquefolia
Ivy, Canary
  see Ivy, Algerian
Ivy, English, 81, *82*, 266, 268
  Hedera helix
Japanese Boxwood
  see Boxwood, Japanese
Japanese Cedar
  see Cryptomeria
Japanese Cherry
  see Cherry, Japanese
Japanese Fatsia
  see Fatsia, Japanese
Japanese Honeysuckle
  see Honeysuckle (vine)
Japanese Jasmine
  see Jasmine, Primrose
Japanese Laurel
  see Aucuba
Japanese Maple
  see Maple, Japanese
Japanese Persimmon
  see Persimmon, Japanese
Japanese Pittosporum
  see Pittosporum, Japanese
Japanese Plum
  see Loquat

Japanese Privet
　see Privet, Japanese
Japanese Quince
　see Quince, Flowering
Japanese Varnish Tree
　see Parasol Tree, Chinese
Japanese Wisteria
　see Wisteria
Japanese Yew
　see Yew, Japanese
Jasmine, Cape
　see Gardenia, Common
Jasmine, Carolina
　see Jasmine, Yellow
Jasmine, Chinese
　see Jasmine, Primrose
Jasmine, Confederate, 215, 266
　Trachelospermum jasminoides
Jasmine, Crape, 206, 254
　Tabernaemontana divaricata
Jasmine, Japanese
　see Jasmine, Primrose
Jasmine, Primrose, 97, 258
　Jasminum mesnyi
Jasmine, Star
　see Jasmine, Confederate
Jasmine, Yellow, 76, *77*, 267
　Gelsemium sempervirens
Jelly Palm
　see Palm, Pinto (Pindo)
Jersey Pine
　see Pine, Poverty
Jerusalem Thorn
　see Thorn, Jerusalem
Jessamine, Yellow
　see Jasmine, Yellow
Jujube, Chinese
　see Jujube, Common
Jujube, Common, *231*, 232, 244
　Zizyphus jujuba

Juniper, Chinese, 98, *99*, 244
　Juniperus chinensis cv. Kaizuka
Juniper, Common, 98, 268
　Juniperus communis
Juniper, Prostrate
　see Juniper, Common
Juniper, Shore, 99, *100*, 266
　Juniperus conferta
Juniper, Virginia
　see Cedar, Eastern Red
Kumquat, Oval, 71, 246
　Fortunella margarita
Kumquat, Round, 70, 258
　Fortunella japonica
Kwanzan Cherry
　see Cherry, Kwanzan
Lace Cape Hydrangea
　see Hydrangea, Lace Cap
Lantana, 104, 258
　Lantana camara
Laurel, Australian
　see Pittosporum, Japanese
Laurel, Black, 78, *79*
　Gordonia lasianthus
Laurel, Cherry, 166, *167*, 244
　Prunus caroliniana
Laurel, Japanese
　see Aucuba
Laurel, Mountain, 100, *102*, 260
　Kalmia latifolia
Laurel Oak
　see Oak, Laurel
Leatherleaf, 123, 260
　Mahonia bealei
Le Conte's Pear
　see Pear, Le Conte
Ligustrum, 107, *108*, 258
　Ligustrum japonicum
Lilac, Summer
　see Butterfly Bush

Lily-flowered Magnolia
   see Magnolia, Lily-flowered
Live Oak
   see Oak, Live
Loblolly Bay
   see Bay, Loblolly
Loblolly Pine
   see Pine, Old Field
Locust, Black, 189, *190*, 238
   Robinia pseudoacacia
Lombardy Poplar
   see Poplar, Lombardy
Longleaf Pine
   see Pine, Longleaf
Loquat, 59, *60*, 246
   Eriobotrya japonica
Love Charm
   see Trumpet Vine, Argentine
Magnolia, Ashe's, 115, *116*, 244
   Magnolia ashei
Magnolia, Bull Bay
   see Magnolia, Southern
Magnolia, Lily-flowered, 119, *120*, 258
   Magnolia liliiflora
Magnolia, Pyramidal, 116, *117*, 248
   Magnolia pyramidata
Magnolia, Saucer, 120, *121*, 248
   Magnolia × *soulangeana*
Magnolia, Silver Bay
   see Bay, Silver (Sweet)
Magnolia, Southern, 117, *118*, 242
   Magnolia grandiflora
Magnolia, Star, 121, *122*, 250
   Magnolia stellata
Magnolia, Tulip
   see Magnolia, Lily-flowered
Maidenhair Tree, 77, *78*, 234
   Ginkgo biloba

Mallow, Common Rose
   see Mallow Rose
Mallow Rose, 82, *83*, 260
   Hibiscus moscheutos
Mandarin Orange
   see Orange, Mandarin
Maple, Japanese, *4*, 5, 246
   Acer palmatum
Maple, Red, 5, *6*, 136, 240
   Acer rubrum (southern form)
March Rose
   see Almond, Flowering
Mexican Palo Verde
   see Thorn, Jerusalem
Mimosa, 9, *10*, 246
   Albizia julibrissin
Mock Orange
   see Orange, Mock
Mockernut Hickory
   see Hickory, Mockernut
Monkey Tree
   see Fir, China
Mountain Laurel
   see Laurel, Mountain
Mulberry, American
   see Beautyberry
Mulberry, French
   see Beautyberry
Mulberry, White, 131, *132*, 242
   Morus alba
Muscadine Grape
   see Grape, Muscadine
Myrtle, Crape, 103, 246
   Lagerstroemia indica
Myrtle, Wax, 132, *133*, 264
   Myrica cerifera
Nandina, 134, 260
   Nandina domestica
Needle Palm
   see Palm, Needle

Oak, Blackjack, 177, *179*, 244
  Quercus marilandica
Oak, Chinese Evergreen, 178, *180*, 238
  Quercus myrsinifolia
Oak, Laurel, 177, *178*, 240
  Quercus laurifolia (*see also* Oak, Water)
Oak, Live, 180, *181*, 240
  Quercus virginiana
Oak, Sand Live, 180, 248
  Quercus virginiana var. *maritima*
Oak, Southern Red, 176, *177*, 242
  Quercus falcata
Oak, Spanish
  see Oak, Southern Red
Oak, Water, 180, *181*, 242
  Quercus nigra (*see also* Oak, Laurel)
Oakleaf Hydrangea
  see Hydrangea, Oakleaf
Old Field Pine
  see Pine, Old Field
Oleander, 135, *136*, 260
  Nerium oleander
Olive, American
  see Devilwood
Olive, Fragrant, 139, *140*, 246
  Osmanthus fragrans
Olive, Tea
  see Olive, Fragrant
Olive, Wild
  see Devilwood
Orange, Mandarin, 45, 260
  Citrus reticulata (*see also* Satsuma)
Orange, Mock, 145, 260
  Philadelphus coronarius, P. inodorus, 146, *147*

Oregon Grape
  see Leatherleaf
Oval Kumquat
  see Kumquat, Oval
Oriental Cherry
  see Cherry, Japanese
Palm, Cabbage, 193, *194*, 223, 238
  Sabal palmetto
Palm, European Fan, 41, *42*, 256
  Chamaerops humilis
Palm, Jelly
  see Palm, Pinto (Pindo)
Palm, Needle, 183, *184*, 260
  Rhapidophyllum hystrix
Palm, Pinto (Pindo), 16, *17*, 248
  Butia capitata
Palm, Porcupine
  see Palm, Needle
Palm, Sago, 54, 229, 262
  Cycas revoluta
Palm, Washington, 193, 222, *223*, 242
  Washingtonia filifera
Palm, Windmill, 216, *217*, 250
  Trachycarpus fortunei
Palmetto, Blue
  see Palm, Needle
Palo Verde, Mexican
  see Thorn, Jerusalem
Paper Plant
  see Fatsia, Japanese
Parasol Tree, Chinese, 67, *68*, 238
  Firmiana simplex
Peach, Flowering, 170
  Prunus persica
Pear, Bradford, 174, 244
  Pyrus calleryana cv. Bradford
Pear, Le Conte, 174
  Pyrus × *lecontei*

Pear, Sand, 174, *175*, 240
  Pyrus pyrifolia
Pecan, 26, *27*, 236
  Carya illinoinensis
Persimmon, Japanese, 56, *57*, 246
  Diospyros kaki
Persimmon, Wild, 56, *58*, 250
  Diospyros virginiana
Phoenix Tree
  see Parasol Tree, Chinese
Pignut Hickory
  see Hickory, Pignut
Pine, Eastern White
  see Pine, White
Pine, Jersey
  see Pine, Poverty
Pine, Loblolly
  see Pine, Old Field
Pine, Longleaf, 156, 234
  Pinus palustris
Pine, Old Field, 158, *159*, 236
  Pinus taeda
Pine, Poverty, 158, *160*, 240
  Pinus virginiana
Pine, Sand, 152, 240
  Pinus clausa
Pine, Scrub
  see Pine, Sand
Pine, Shortleaf
  see Pine, Shortleaf Yellow
Pine, Shortleaf Yellow, 153, 236
  Pinus echinata
Pine, Slash, 154, 236
  Pinus elliottii
Pine, Southern
  see Pine, Longleaf
Pine, Spruce, 155, 236
  Pinus glabra
Pine, Swamp
  see Pine, Slash
Pine, Virginia
  see Pine, Poverty
Pine, White, 157, 236
  Pinus strobus
Pine, Yellow
  see Pine, Longleaf
Pineapple Guava
  see Feijoa
Pinto (Pindo) Palm
  see Palm, Pinto
Pinwheel Flower
  see Jasmine, Crape
Pinxter Flower
  see Honeysuckle (bush)
Pittosporum, Japanese, 160, *161*, 258
  Pittosporum tobira
Plum, Chickasaw, 166, 244
  Prunus angustifolia
Plum, Japanese
  see Loquat
Plum-leaved Azalea
  see Azalea, Plum-leaved
Pomegranate, 171, *172*, 260
  Punica granatum
Pond Cypress
  see Cypress, Pond
Popcorn Tree, 197, *198*, 248
  Sapium sebiferum
Poplar, Lombardy, 164, *165*, 240
  Populus nigra cv. Italica
Poplar, Yellow
  see Tulip Tree
Porcupine Palm
  see Palm, Needle
Poverty Pine
  see Pine, Poverty
Prickly Pear, 137, *138*, 260
  Opuntia ficus-indica
Primrose Jasmine
  see Jasmine, Primrose

Princess Tree, 144, *145*, 240
 Paulownia tomentosa
Privet, Chinese
 see Privet, Glossy
Privet, Common, 108, *109*, 254
 Ligustrum sinense
Privet, Glossy, 108, 246
 Ligustrum lucidum
Privet, Japanese
 see Ligustrum
Privet, Wax Leaf
 see Ligustrum
Prostrate Juniper
 see Juniper, Common
Pyramidal Magnolia
 see Magnolia, Pyramidal
Quince, Flowering, 36, *37*, 256
 Chaenomeles speciosa
Quince, Japanese
 see Quince, Flowering
Red Birch
 see Birch, River
Red Buckeye
 see Buckeye, Red
Red Cedar, Eastern
 see Cedar, Eastern Red
Red Maple
 see Maple, Red
Red Tip (Top), 146, *148*, 248
 Photinia × *fraseri*
Redbud, 35, *36*, 248
 Cercis canadensis
Reed, Giant, 12, 256
 Arundo donax
Rice Paper, Formosa
 see Fatsia, Japanese
Rice Paper Plant, 212, *213*, 262
 Tetrapanax papyriferus
River Birch
 see Birch, River

Rose, 190
 Rosa (all types)
Rose, Confederate, 84, 254
 Hibiscus mutabilis
Rose, March
 see Almond, Flowering
Rose Mallow, Common
 see Mallow Rose
Rose of China
 see Hibiscus, Chinese
Rose of Sharon
 see Althea
Round Kumquat
 see Kumquat, Round
Sage, Texas, 106, 264
 Leucophyllum frutescens
Sage, Yellow
 see Lantana
Sago Palm
 see Palm, Sago
Sand Live Oak
 see Oak, Sand Live
Sand Pear
 see Pear, Sand
Sand Pine
 see Pine, Sand
Sasanqua
 see Camellia, Sasanqua
Sassafras, 199, *200*, 242
 Sassafras albidum
Satsuma, 45, 248
 Citrus reticulata (see also Orange, Mandarin)
Saucer Magnolia
 see Magnolia, Saucer
Scarlet Wisteria Tree, 200, *201*, 262
 Sesbania punicea
Scrub Pine
 see Pine, Sand

Scuppernong
  see Grape, Muscadine
Seminole Bread
  see Coontie
Shore Juniper
  see Juniper, Shore
Shortleaf Pine
  see Pine, Shortleaf Yellow
Shortleaved Yellow Pine
  see Pine, Shortleaf Yellow
Silk Tree
  see Mimosa
Silky Camellia
  see Camellia, Silky
Silver Bay
  see Bay, Silver
Silver-bell Tree, 80, *81*, 248
  Halesia diptera
Silver Dollar Tree, 61, 242
  Eucalyptus polyanthemos (?)
Silver Thorn, 57, *59*, 262
  Elaeagnus pungens cv. Simonii
Slash Pine
  see Pine, Slash
Sloe, 165, 248
  Prunus americana or *alleghaniensis*
Smoke Tree, 49, 248
  Cotinus coggygria
Snowbell, 205, *207*, 262
  Styrax americanus
Snowdrop Tree, 80, 250
  Halesia carolina
Sour Gum, 136, 242
  Nyssa sylvatica
Sourwood, 140, *141*, 242
  Oxydendron arboreum
Southern Magnolia
  see Magnolia, Southern

Southern Pine
  see Pine, Longleaf
Southern Red Oak
  see Oak, Southern Red
Southern Yew
  see Yew, Japanese
Spanish Bayonet
  see Bayonet, Spanish
Spanish Oak
  see Oak, Southern Red
Sparkleberry, 218, *219*, 262
  Vaccinium arboreum
Spindle Tree, 63, *64*, 262
  Euonymus japonicus
Spruce Pine
  see Pine, Spruce
Star Jasmine
  see Jasmine, Confederate
Star Magnolia
  see Magnolia, Star
Strawberry Bush, 62, *63*, 264
  Euonymus americanus
Summer Lilac
  see Butterfly Bush
Swamp Blueberry
  see Huckleberry, Highbush
Swamp Pine
  see Pine, Slash
Sweet Bay
  see Bay, Sweet
Sweet Gum, 110, 236
  Liquidambar styraciflua
Sweet Shrub, 20, *21*, 264
  Calycanthus floridus
Sycamore, Eastern, 162, 236
  Platanus occidentalis
Taiwan Cherry
  see Cherry, Taiwan
Tallow Tree, Chinese
  see Popcorn Tree

Tea Olive
   see Olive, Fragrant
Ternstroemia, 211, *212*, 250
   Ternstroemia gymnanthera
Texas Sage
   see Sage, Texas
Thorn
   see Haw
Thorn, Jerusalem, 142, 246
   Parkinsonia aculeata
Thorn, Silver
   see Silver Thorn
Thorn Apple
   see Haw
Thorny Elaeagnus
   see Silver Thorn
Trumpet Creeper
   see Creeper, Trumpet
Trumpet Flower, Evening
   see Jasmine, Yellow
Trumpet Honeysuckle
   see Honeysuckle, Trumpet
Trumpet Vine
   see Creeper, Trumpet
Trumpet Vine, Argentine, 46, *47*, 266
   Clytostoma callistegioides
Tulip Magnolia
   see Magnolia, Lily-flowered
Tulip Tree, 111, *112*, 236
   Liriodendron tulipifera
Tung Oil Tree, 11, 250
   Aleurites fordii
Tupelo
   see Sour Gum
Turk's Cap Hibiscus
   see Hibiscus, Turk's Cap
Varnish Tree, Japanese
   see Parasol Tree, Chinese

Virginia Creeper
   see Creeper, Virginia
Virginia Juniper
   see Cedar, Eastern Red
Virginia Pine
   see Pine, Poverty
Wahoo
   see Burning Bush
Washington Palm
   see Palm, Washington
Water Oak
   see Oak, Water
Wax Bayberry
   see Myrtle, Wax
Wax-leaf Privet
   see Privet, Japanese
Wax Myrtle
   see Myrtle, Wax
Weeping Willow
   see Willow, Weeping
Weigela, 224, 264
   Weigela florida
White Ash
   see Ash, White
White Cedar
   see Cedar, White
White Mulberry
   see Mulberry, White
White Pine
   see Pine, White
Whortleberry
   see Huckleberry, Highbush
Wild Cassava
   see Cassava, Wild
Wild Olive
   see Devilwood
Wild Persimmon
   see Persimmon, Wild
Willow, Black, 196, 238
   Salix nigra

Willow, Weeping, 196, 250
  Salix babylonica
Windmill Palm
  see Palm, Windmill
Winter Creeper
  see Creeper, Winter
Wintersweet, 41, 264
  Chimonanthus praecox
Wisteria (or Japanese Wisteria), 225, 267
  Wisteria floribunda and/or *W. sinensis*, 225, 226
Woodbine
  see Creeper, Virginia
Worm Tree
  see Catalpa
Yaupon, 94, 95, 264
  Ilex vomitoria

Yellow Jasmine
  see Jasmine, Yellow
Yellow Jessamine
  see Jasmine, Yellow
Yellow Pine
  see Pine, Longleaf
Yellow Poplar
  see Tulip Tree
Yellow Sage
  see Sage, Yellow
Yew, Florida, 210, *211*, 256
  Taxus floridana
Yew, Japanese, 163, *164*, 240
  Podocarpus macrophyllus
Yew, Southern
  see Yew, Japanese

Index of Scientific Names

Genera, Families, Species

*An asterisk before the species name indicates a native species in Zone 8. Illustrations not on same page as text are designated by italic page numbers.

Abelia: Caprifoliaceae, 3
   Abelia × *grandiflora* (Andre) Rehd., *3*, 252
Acer: Aceraceae, 4
   Acer palmatum Thunb., 4, 5, 246
   *A. rubrum L., 5, *6*, 240
Aesculus: Hippocastanaceae, 5
   *Aesculus pavia L., 6, *7*, 260
Agave: Agavaceae, 7
   Agave americana L., *8*, 254
Albizia: Leguminosae, 9
   Albizia julibrissin Durazz, 9, *10*, 246
Aleurites: Euphorbiaceae, 9
   Aleurites fordii Hemsl., 11, 250
Arundo: Gramineae, 11
   Arundo donax L., 12, 256
Aucuba: Cornaceae, 13
   Aucuba japonica Thunb., 13, *14*, 252

Betula: Betulaceae, 13
   *Betula nigra L., 14, *15*, 240
Buddleja: Loganiaceae, 15
   Buddleja davidii Franch., 16, 252
Butia: Palmae, 16
   Butia capitata (Mart.) Becc., 16, *17*, 248
Buxus: Buxaceae, 17
   Buxus microphylla Sieb. & Zucc., var. *japonica*, 17, 252
Callicarpa: Verbenaceae, 18
   *Callicarpa americana L., 18, *19*, 252
Callistemon: Myrtaceae, 18
   Callistemon citrinus (Curtis) Skeels or *C. viminalis* (Sol. ex Gaertn.) G. Don, 19, *20*, 254
Calycanthus: Calycanthaceae, 19

Calycanthaceae (*continued*)
   *Calycanthus floridus L., 20, *21*, 264
Camellia: Theaceae, 21
   Camellia japonica L., 22, 252
   C. sasanqua Thunb., 23, 252
Campsis: Bignoniaceae, 24
   *Campsis radicans (L.) Seem. ex Bur., 24, 266
Carya: Juglandaceae, 25
   *Carya glabra (Mill.) Sweet, 25, 26, 236
   *C. illinoinensis (Wangenh.) K. Koch, 26, *27*, 236
   *C. tomentosa (Lam.) Nutt., 27, 235
Cassia: Leguminosae, 28
   Cassia corymbosa Lam., 28, *29*, 254
   C. coluteoides Call., 28, *30*, 254
Castanea: Fagaceae, 29
   Castanea mollissima Bl., 30, *31*, 238
   *Castanea pumila (L.) Mill., 32, 244
Catalpa: Bignoniaceae, 33
   *Catalpa bignonioides Walt., 33, *34*, 242
Cedrus: Pinaceae, 33
   Cedrus deodara (Roxb.) Loud., 34, *35*, 234
Cercis: Leguminosae, 35
   *Cercis canadensis L., 35, *36*, 248
Chaenomeles: Rosaceae, 36
   Chaenomeles speciosa (Sweet) Nakai, 37, 256
Chamaecyparis: Cupressaceae, 38
   *Chamaecyparis thyoides (L.) B.S.P., 38, *39*, 236
   *C. thyoides cv. Ericoides, 39, *40*, 244

Chamaerops: Palmae, 41
   Chamaerops humilis L., 41, *42*, 256
Chimonanthus: Calycanthaceae, 41
   Chimonanthus praecox (L.) Link, 41, 264
Chionanthus: Oleaceae, 42
   *Chionanthus virginicus L., 43, 246
Cinnamomum: Lauraceae, 43
   Cinnamomum camphora (L.) Nees & Eberm., 44, 234
Citrus: Rutaceae, 45
   Citrus reticulata Blanco, 45, 248, 260
Clytostoma: Bignoniaceae, 46
   Clytostoma callistegioides (Cham.) Bur., 46, *47*, 266
Cornus: Cornaceae, 46
   *Cornus florida L., 47, *48*, 246
Cotinus: Anacardiaceae, 48
   Cotinus coggygria Scop., 49, 248
Crataegus: Rosaceae, 50
   *Crataegus uniflora Meunch., 50, *51*, 246
Cryptomeria: Taxodiaceae, 51
   Cryptomeria japonica (L. f.) D. Don, 51, *52*, 234
Cunninghamia: Taxodiaceae, 52
   Cunninghamia lanceolata (Lamb.) Hook. F., 53, 234
Cycas: Cycadaceae, 53
   Cycas revoluta Thunb., 54, 229, 264
Deutzia: Saxifragaceae, 54
   Deutzia scabra cv. Candidissima Thunb., 55, 254
Diospyros: Ebenaceae, 56
   Diospyros kaki L. f., 56, *57*, 246
   *D. virginiana L., 56, *58*, 250

Elaeagnus: Elaeagnaceae, 57
  Elaeagnus pungens cv. Simonii
    Thunb., 57, 59, 262
Eriobotrya: Rosaceae, 58
  Eriobotrya japonica (Thunb.)
    Lindl., 59, 60, 246
Eucalyptus: Myrtaceae, 60
  Eucalyptus polyanthemos (?)
    Schaur, 61, 242
Euonymus: Celastraceae, 62
  *Euonymus americanus L., 62,
    63, 254
  *E. atropurpureus Jacq., 62, 252
  E. fortunei (Turcz.) Hand.-
    Mazz., 62, 258
  E. japonicus Thunb., 63, 64, 262
Fatsia: Araliaceae, 64
  Fatsia japonica (Thunb.) Decn. &
    Planch., 64, 65, 256
Feijoa: Myrtaceae, 65
  Feijoa sellowiana (Berg) Berg, 66,
    256
Firmiana: Sterculiaceae, 67
  Firmiana simplex W. F. Wright.,
    67, 68, 238
Forsythia: Oleaceae, 67
  Forsythia × intermedia Zabel,
    68, 69, 258
Fortunella: Rutaceae, 69
  Fortunella japonica (Thunb.)
    Swingle, 70, 258
  F. margarita (Lour.) Swingle,
    71, 246
Franklinia: Theaceae, 71
  *Franklinia alatamaha Marsh.,
    71, 246
Fraxinus: Oleaceae, 72
  *Fraxinus americana L., 73, 236
Gardenia: Rubiaceae, 74
  Gardenia jasminoides Ellis, 74,
    75, 256

G. jasminoides Ellis cv. Radicans,
    75, 254
Gelsemium: Loganiaceae, 76
  *Gelsemium sempervirens (Lait.
    f.), 76, 77, 267
Ginkgo: Ginkgoaceae, 76
  Ginkgo biloba L., 77, 78, 234
Gordonia: Theaceae, 78
  *Gordonia lasianthus (L.) Ellis,
    78, 79
Halesia: Styracaceae, 79
  *Halesia carolina L., 80, 250
  *H. diptera Ellis, 80, 81, 248
Hedera: Araliaceae, 80
  Hedera canariensis Willd., 81,
    266
  H. helix L., 81, 82, 266, 268
Hibiscus: Malvaceae, 82
  *Hibiscus moscheutos L., 82, 83,
    260
  H. mutabilis L., 84, 254
  H. rosa-sinensis L., 85, 254
  H. syriacus L., 86, 252
Hydrangea: Saxifragaceae, 87
  Hydrangea macrophylla (Thunb.)
    Ser. var. *macrophylla* c.v. Hor-
    tensia, 87, 88, 258
  H. macrophylla var. *macrophylla*
    cv. Lace Cap, 88, 89, 258
  *H. quercifolia Bartr., 89, 90,
    260
Ilex: Aquifoliaceae, 90
  Ilex × attenuata cv. East Palatka,
    91, 238
  I. cornuta Lindl. & Paxton, 92
  I. cornuta Lindl. & Paxton cv.
    Burfordii, 92, 93, 244
  I. cornuta cv. Rotunda, 92
  *I. opaca Ait. f, 93, 94, 238
  *I. vomitoria Ait., 94, 95, 264

Illicium: Illiciaceae, 95
Illicium floridanum Ellis, 96, 256
Jasminum: Oleaceae, 97
Jasminum mesnyi Hance, 97, 258
Juniperus: Cupressaceae, 98
Juniperus chinensis L. cv. Kaizuka, 98, *99*, 244
J. communis L., 98, 268
J. conferta Parl., 99, *100*, 268
**J. virginiana* L., 99, *101*, 238
Kalmia: Ericaceae, 100
Kalmia latifolia L., 100, *102*, 260
Lagerstroemia: Lythraceae, 102
Lagerstroemia indica L., 103, 246
Lantana: Verbenaceae, 104
Lantana camara L., 104, 258
Leucophyllum: Scrophulariaceae, 105
Leucophyllum frutescens (Berl.) Johnst., 105, *106*, 264
Leucothoe: Ericaceae, 106
**Leucothoe populifolia* (Lam.) Dipp., 107, 256
Ligustrum: Oleaceae, 107
Ligustrum japonicum Thunb., 107, *108*, 258
L. lucidum Ait. f., 108, 246
L. sinense Lour., 108, *109*, 254
Liquidambar: Hamamelidaceae, 109
**Liquidambar styraciflua* L., 110, 236
Liriodendron: Magnoliaceae, 111
**Liriodendron tulipifera* L., 111, *112*, 236
Lonicera: Caprifoliaceae, 112
Lonicera japonica Thunb., 113, 266

L. sempervirens L., 114, 266
Magnolia: Magnoliaceae, 114
M. × *soulangeana* Soul.-Bod., 120, *121*, 248
**M. ashei* Weatherby, 115, *116*, 244
**M. grandiflora* L., 117, *118*, 242
M. liliiflora Desr., 119, *120*, 258
**M. pyramidata* Bartr. ex Pursh, 116, *117*, 248
M. stellata (Sieb. & Zucc.) Maxim., 121, *122*, 250
**M. virginiana* var. *australis* Sarg., 118, *119*, 242
Mahonia: Berberidaceae, 122
Mahonia bealei (Fortune) Carr., 123, 260
Malus: Rosaceae, 124
Malus sylvestris Mill., 124, *125*, 244
M. (several species) = crabapple, 124, *126*, 244
Malvaviscus: Malvaceae, 125
Malvaviscus arboreus var. *mexicanus* Schlechtend., 126, *127*, 264
Manihot: Euphorbiaceae, 127
Manihot grahamii Hook., 127, *128*, 250
Melia: Meliaceae, 129
Melia azedarach L., 129, 238
Michelia: Magnoliaceae, 130
Michelia figo (Lour.) Spreng., 130, *131*, 252
Morus: Moraceae, 130
Morus alba L., 131, *132*, 242
Myrica: Myricaceae, 132
**Myrica cerifera* L., 132, *133*, 264

Nandina: Berberidaceae, 133
Nandina domestica Thunb., 134, 260
Nerium: Apocynaceae, 135
Nerium oleander L., 135, *136*, 260
Nyssa: Nyssaceae, 135
*Nyssa sylvatica Marsh., *136*, 242
Opuntia: Cactaceae, 137
Opuntia ficus-indica (L.) Mill., 137, *138*, 260
Osmanthus: Oleaceae, 138
**Osmanthus americanus* (L.) Gray, 138, *139*, 268
O. fragrans Lour., 139, *140*, 246
Oxydendron: Ericaceae, 140
**Oxydendron arboreum* (L.) DC, 140, *141*, 242
Parkinsonia: Leguminosae, 141
Parkinsonia aculeata L., 142, 246
Parthenocissus: Vitaceae, 143
**Parthenocissus quinquefolia* (L.) Planch., 143, *144*, 267
Paulownia: Bignoniaceae, 143
Paulownia tomentosa (Thunb.) Steud., 144, *145*, 240
Philadelphus: Saxifragaceae, 145
Philadelphus coronarius L., 146, 260
P. inodorus L., 146, *147*
Photinia: Rosaceae, 146
Photinia × *fraseri* W. J. Dress., 146, *148*, 248
P. serrulata Lindl., 147, *149*, 248
Phyllostachys: Gramineae, 148
Phyllostachys aurea A. & C. Riv.(?), 149, *150*, 252
Pinus: Pinaceae, 150
*Pinus clausa (Chapm. ex Engelm.) Vasey ex Sarg., 152, 240
**P. echinata* Mill., 153, 236
**P. elliottii* Engelm., 154, 236
**P. glabra* Walt., 155, 236
**P. palustris* Mill., 156, 234
P. strobus L., 157, 236
**P. taeda* L., 158, *159*, 236
P. virginiana Mill., 158, *160*, 240
Pittosporum: Pittosporaceae, 159
Pittosporum tobira (Willd.) Ait., 160, *161*, 258
Platanus: Platanaceae, 161
**Platanus occidentalis* L., 162, 236
Podocarpus: Podocarpaceae, 163
Podocarpus macrophyllus (Thunb.) D. Don, 163, *164*, 240
Populus: Salicaceae, 163
Populus nigra cv. Italica L., 164, *165*, 240
Prunus: Rosaceae, 165
*Prunus americana Marsh. (or *P. alleghaniensis*) Porter, 165, 248
**P. angustifolia* Marsh., 166, 244
P. campanulata Maxim., 167, *168*, 250
**P. caroliniana* Ait., 166, *167*, 244
P. glandulosa Thunb. cv. Sinensis, 169, 256
P. persica (L.) Batch, 170
P. serrulata Lindl. c.v. Kwanzan, 170, 240
Punica: Punicaceae, 170
Punica granatum L., 171, *172*, 260
Pyracantha: Rosaceae, 171
Pyracantha coccinea Roem., 172, *173*, 256

Pyracantha (*continued*)
P. *crenulata* (D. Don) Roem., 173, 256
Pyrus: Rosaceae, 173
Pyrus calleryana Decene., 174, 244
P. *pyrifolia* (Burm.) Nakai, 174, *175*, 244
P. × *lecontei* Rehd., 174
Quercus: Fagaceae, 174
Quercus falcata Michx., 176, *177*, 242
Q. laurifolia Michx., 177, *178*
Q. marilandica Muenchh., 177, *179*, 244
Q. *myrsinifolia* Bl., 178, *180*, 238
Q. nigra L., 180, *181*, 242
Q. virginiana Mill., 180, *181*, 240
Q. virginiana var. *maritima* (Chapm.) Sarg., 180, 248
Rhaphiolepis: Rosaceae, 180
Rhaphiolepis indica (L.) Lindl., 182, *183*, 258
Rhapidophyllum: Palmae, 183
Rhapidophyllum hystrix (Pursh) H. Wendl. & Drude, 183, *184*, 260
Rhododendron: Ericaceae, 185
Rhododendron alabamense Rehd., 186, *187*, 252
R. austrinum (Small) Rehd., 187, *188*, 256
R. canescens (Michx.) Sweet, 188, 258
R. chapmanii A. Gray, 188, 254
R. prunifolium (Small) Mallais, 189, 260
Robinia: Leguminosae, 189

Robinia pseudoacacia L., 189, *190*, 238
Rosa: Rosaceae, 190
Rosa species
Sabal: Palmae, 192
Sabal palmetto (Walt.) Lodd. ex Schult. & Schult. f., 193, *194*, 223, 238
Salix: Salicaceae, 195
Salix babylonica L., 196, 250
S. nigra. L., 196, 238
Sapium: Euphorbiaceae, 197
Sapium sebiferum (L.) Roxb., 197, *198*, 248
Sassafras: Lauraceae, 199
Sassafras albidum (Nutt.) Nees, 199, *200*, 242
Sesbania: Leguminosae, 199
Sesbania punicea (Cav.) Benth., 200, *201*, 262
Spiraea: Rosaceae, 201
Spiraea cantoniensis Laur., 201, *202*, 262
S. *prunifolia* Sieb. & Zucc., 202, *203*, 262
S. *thunbergii* Sieb. ex Bl., 202, *204*, 262
S. × *bumalda* Burv., 203, *205*, 262
Stewartia: Theaceae, 203
Stewartia malacodendron L., 204, *206*, 262
Styrax: Styraceae, 204
Styrax americanus Lam., 205, *207*, 262
Tabernaemontana: Apocynaceae, 206
Tabernaemontana divaricata (L.) R. Br., 206, 254

Taxodium: Taxodiaceae, 207
　Taxodium ascendens Brongn.,
　　208, 240
　T. distichum (L.) R. Rich., 209,
　　234
Taxus: Taxaceae, 210
　Taxus floridana Nutt. ex
　　Chapm., 210, *211*, 256
Ternstroemia: Theaceae, 211
　Ternstroemia gymnanthera (Wight
　　& Arn.) Sprague, 212, 250
Tetrapanax: Araliaceae, 212
　Tetrapanax papyriferus (Hook.)
　　K. Koch, 212, *213*, 262
Thuja: Cupressaceae, 213
　Thuja occidentalis L., 213, *214*,
　　238
Trachelospermum: Apocynaceae,
　　214
　Trachelospermum jasminoides
　　(Lindl.) Lam., 215, 266
Trachycarpus: Palmae, 216
　Trachycarpus fortunei (Hook.)
　　Wendl., 216, *217*, 250
Tsuga: Pinaceae, 216
　Tsuga caroliniana Engelm., 217,
　　218, 238
Vaccinium: Ericaceae, 218

*Vaccinium arboreum Marsh.,
　　218, *219*, 262
*V. corymbosum L., 219, *220*,
　　258
Vitis: Vitaceae, 220
　*Vitis rotundifolia Michx., 221,
　　222, 266
Washingtonia: Palmae, 222
　Washingtonia filifera (Linden ex
　　Andre) H. A. Wendland, 193,
　　222, *223*, 242
Weigela: Caprifoliaceae, 223
　Weigela florida (Bunge) A. DC.,
　　224, 264
Wisteria: Leguminosae, 225
　Wisteria floribunda (Willd.) DC.,
　　225, 267
　W. sinensis (Sims) Sweet, 226
Yucca: Agavaceae, 227
　*Yucca aloifolia L., 227, *228*, 262
　*Y. flaccida Haw., 227, *229*, 252
Zamia: Zamiaceae, 229
　*Zamia pumila L., 230, 254
Zizyphus: Rhamnaceae, 231
　Zizyphus jujuba Mill, *231*, 232,
　　244

Index of Families, Genera, and Species

Aceraceae, 4
  Acer palmatum, 4, 5, 246
  A. rubrum, 5, 6, 240
Agavaceae, 7, 227
  Agave americana, 8, 254
  Yucca aloifolia, 227, 228, 262
  Y. flaccida, 227, 229, 252
Anacardiaceae, 48
  Cotinus coggygria, 49, 248
Apocynaceae, 135, 206, 214
  Nerium oleander, 135, *136*, 260
  Tabernaemontana divaricata, 206, 254
  Trachelospermum jasminoides, 215, 266
Aquifoliaceae, 90
  Ilex × *attenuata* cv. East Palatka, 91, 238
  Ilex cornuta, 92
  I. cornuta cv. Burfordi, 92, *93*, 244
  I. cornuta cv. Rotunda, 92

I. opaca, 93, *94*, 238
I. vomitoria, 94, *95*, 264
Araliaceae, 64, 80, 212
  Fatsia japonica, 64, *65*, 256
  Hedera canariensis, 81, 266
  H. helix, 81, *82*, 266, 268
  Tetrapanax papyriferus, 212, *213*, 262
Berberidaceae, 122, 133
  Mahonia bealei, 123, 260
  Nandina domestica, 134, 260
Betulaceae, 13
  Betula nigra, 14, *15*, 240
Bignoniaceae, 24, 33, 46, 143
  Campsis radicans, 24, 266
  Catalpa bignonioides, 33, *34*, 242
  Clytostoma callistegioides, 46, *47*, 266
  Paulownia tomentosa, 144, *145*, 240
Buxaceae, 17

Buxus microphylla var. *japonica*, 17, 252
Cactaceae, 137
　Opuntia ficus-indica, 137, *138*, 260
Calycanthaceae, 19, 41
　Calycanthus floridus, 20, *21*, 264
　Chimonanthus praecox, 41, 264
Caprifoliaceae, 3, 112, 223
　Abelia × *grandiflora*, 3, 252
　Lonicera japonica, 113, 266
　L. sempervirens, 114, 266
　Weigela florida, 224, 264
Celastraceae, 62
　Euonymus americanus, 62, *63*, 254
　E. atropurpureus, 62, 252
　E. fortunei, 62, 258
　E. japonicus, 63, *64*, 262
Cornaceae, 13, 46
　Aucuba japonica, 13, *14*, 252
　Cornus florida, 47, *48*, 246
Cupressaceae, 38, 98
　Chamaecyparis thyoides, 38, *39*, 236
　C. thyoides cv. Ericoides, 39, *40*, 244
　Juniperus chinensis cv. Kaizuka, 98, *99*, 244
　J. communis, 98, 268
　J. conferta, 99, *100*, 268
　J. virginiana, 99, *101*, 238
Cupressaceae, 213
　Thuja occidentalis, 213, *214*, 238
Cycadaceae, 53
　Cycas revoluta, 54, 262
Ebenaceae, 56
　Diospyrus kaki, 56, *57*, 246
　D. virginiana, 56, *58*, 250
Elaeagnaceae, 57

Elaeagnus pungens cv. Simonii, 57, *59*, 262
Ericaceae, 100, 106, 140, 185, 218
　Kalmia latifolia, 100, *102*, 260
　Leucothoe populifolia, 107, 256
　Oxydendron arboreum, 140, *141*, 242
　Rhododendron alabamense, 186, *187*, 252
　R. austrinum, 187, *188*, 256
　R. canescens, 188, 258
　R. chapmanii, 188, 254
　R. prunifolium, 189, 260
　Vaccinium arboreum, 218, *219*, 262
　V. corymbosum, 219, *220*, 258
Euphorbiaceae, 9, 127, 197
　Aleurites fordii, 11, 250
　Manihot grahamii, 127, *128*, 250
　Sapium sebiferum, 197, *198*, 248
Fagaceae, 29, 174
　Castanea mollisima, 30, *31*, 238
　C. pumila, *32*, 244
　Quercus falcata, 176, *177*, 242
　Q. laurifolia, 177, *178*
　Q. marilandica, 177, *179*, 244
　Q. myrsinifolia, 178, *180*, 238
　Q. nigra, 180, *181*, 242
　Q. virginiana, 180, *181*, 240
　Q. virginiana var. *maritima*, 180, 248
Ginkgoaceae, 76
　Ginkgo biloba, 77, *78*, 234
Gramineae, 11, 148
　Arundo donax, 12, 256
　Phyllostachys aurea, 149, *150*, 252
Hamamelidaceae, 109
　Liquidambar styraciflua, 110, 236
Hippocastanaceae, 109
　Aesculus pavia, 6, *7*, 260

Illiciaceae, 95
  Illicium floridanum, 96, 256
Juglandaceae, 25
  Carya glabra, 25, 26, 236
  C. illinoinensis, 25, 26, 236
  C. tomentosa, 27, 235
Lauraceae, 43, 199
  Cinnamomum camphora, 44, 234
  Sassafras albidum, 199, 200, 242
Leguminosae, 9, 28, 35, 141, 189, 199, 225
  Albizia julibrissin, 9, 10, 246
  Cassia corymbosa, 28, 29, 254
  C. coluteoides, 28, 30, 254
  Cercis canadensis, 35, 36, 248
  Parkinsonia aculeata, 141, 142, 246
  Robinia pseudoacacia, 189, 190, 238
  Sesbania punicea, 200, 201, 262
  Wisteria floribunda, 225, 267
  W. sinensis, 226
Loganiaceae, 15, 76
  Buddleja davidii, 16, 252
  Gelsemium sempervirens, 76, 77, 267
Lythraceae, 102
  Lagerstroemia indica, 103, 246
Magnoliaceae, 111, 114, 130
  Liriodendron tulipifera, 111, 112, 236
  Magnolia ashei, 115, 116, 244
  M. grandiflora, 117, 118, 242
  M. liliiflora, 119, 120, 258
  M. pyramidata, 116, 117, 248
  M. × soulangeana, 120, 121, 248
  M. stellata, 121, 122, 250
  M. virginiana var. *australis*, 118, 119, 242
  Michelia figo, 130, 131, 252

Malvaceae, 82, 125
  Hibiscus moscheutos, 82, 83, 260
  H. mutabilis, 84, 254
  H. rosa-sinensis, 85, 254
  H. syriacus, 86, 252
  Malvaviscus arboreus var. *mexicanus*, 126, 127, 264
Meliaceae, 129
  Melia azedarach, 129, 238
Moraceae, 130
  Morus alba, 131, 132, 242
Myricaceae, 132
  Myrica cerifera, 132, 133, 264
Myrtaceae, 18, 60, 65
  Callistemon citrinus (or *C. viminalis*), 19, 20, 254
  Eucalyptus polyanthemos (?), 61, 242
  Feijoa sellowiana, 66, 256
Nyssaceae, 135
  Nyssa sylvatica, 136, 242
Oleaceae, 42, 67, 72, 97, 107, 138
  Chionanthus virginicus, 43, 246
  Forsythia × *intermedia*, 68, 69, 258
  Fraxinus americana, 73, 236
  Jasminum mesnyi, 97, 258
  Ligustrum japonicum, 107, 108, 258
  L. lucidum, 108, 246
  L. sinense, 108, 109, 254
  Osmanthus americanus, 138, 139, 268
  O. fragrans, 139, 140, 246
Palmae, 16, 41, 183, 192, 216, 222
  Butia capitata, 16, 17, 248
  Chamaerops humilis, 41, 42, 256
  Rhapidophyllum hystrix, 183, 184, 260
  Sabal palmetto, 193, 194, 238

Trachycarpus fortunei, 216, *217*, 260
Washingtonia filifera, 193, *194*, 238
Pinaceae, 33, 150, 216
   Cedrus deodara, 34, *35*, 234
   Pinus clausa, 152, 240
   P. echinata, 153, 236
   P. elliottii, 154, 236
   P. glabra, 155, 236
   P. palustris, 156, 234
   P. strobus, 157, 236
   P. taeda, 158, *159*, 236
   P. virginiana, 158, *160*, 240
   Tsuga caroliniana, 217, *218*, 238
Pittosporaceae, 159
   Pittosporum tobira, 160, *161*, 258
Platanaceae, 161
   Platanus occidentalis, 162, 236
Podocarpaceae, 163
   Podocarpus macrophyllus, 163, *164*, 240
Punicaceae, 170
   Punica granatum, 171, *172*, 260
Rhamnaceae, 231
   Zizyphus jujuba, 231, *232*, 244
Rosaceae, 36, 50, 58, 124, 146, 165, 171, 173, 180, 190, 201
   Chaenomeles speciosa, 37, 256
   Crataegus uniflora, 50, *51*, 246
   Eriobotrya japonica, 59, *60*, 246
   Malus sylvestris, 124, *125*, 244
   Malus spp. (crabapples), 124, *126*, 244
   Photinia × *fraseri*, 146, *148*, 248
   P. serrulata, 147, *149*, 248
   Prunus americana (or *alleghaniensis*), 165, 248
   P. angustifolia, 166, 244
   P. campanulata, 167, *168*, 250
   P. caroliniana, 166, *167*, 244
   P. glandulosa cv. Sinensis, 169, 256
   P. persica, 170
   P. serrulata cv. Kwanzan, 170, 240
   Pyracantha coccinia, 172, *173*, 256
   P. crenulata, 173, 254
   Pyrus calleryana, 174, 244
   P. pyrifolia, 174, *175*, 240 (cf. *P.* × *lecontei*)
   Rhaphiolepis indica, 182, *183*, 258
   Rosa (no species given), 190
   Spiraea cantoniensis, 201, *202*, 262
   S. prunifolia, 202, *203*, 262
   S. thunbergii, 202, *204*, 262
   S. × *bumalda*, 203, *205*, 262
Rubiaceae, 74
   Gardenia jasminoides, 74, *75*, 256
   G. jasminoides cv. Radicans, 75, 254
Rutaceae, 45, 69
   Citrus reticulata, 45, 248, 260
   Fortunella japonica, 70, 258
   F. margarita, 71, 246
Salicaceae, 163, 195
   Populus nigra cv. Italica, 164, 165, 240
   Salix babylonica, 196, 250
   S. nigra, 196, 238
Saxifragaceae, 54, 87, 145
   Deutzia scabra cv. Candidissima, 55, 254
   Hydrangea macrophylla var. *macrophylla* cv. Hortensia, 87, *88*, 258
   H. macrophylla var. *macrophylla* cv. Lace Cap, 88, *89*, 258
   H. quercifolia, 89, *90*, 260

Saxifragaceae (*continued*)
　Philadelphus coronarius, 146, 260
　P. inodorus, 146, *147*
Scrophulariaceae, 105
　Leucophyllum frutescens, 105, *106*, 264
Sterculiaceae, 67
　Firmiana simplex, 67, *68*, 238
Styracaceae, 79, 204
　Halesia carolina, 80, 248
　H. diptera, 80, *81*, 248
　Styrax americanus, 205, *207*, 262
Taxaceae, 210
　Taxus floridana, 210, *211*, 256
Taxodiaceae, 51, *52*, 207
　Cryptomeria japonica, 51, *52*, 234
　Cunninghamia lanceolata, 53, 234
　Taxodium ascendens, 208, 246
　T. distichum, 209, 234

Theaceae, 21, 71, 78, 203, 211
　Camellia japonica, 22, 252
　C. sasanqua, 23, 252
　Franklinia alatamaha, 71, 246
　Gordonia lasianthus, 78, *79*
　Stewartia malacodendron, 204, *206*, 262
　Ternstroemia gymnanthera, 212, 250
Verbenaceae, 18, 104
　Callicarpa americana, 18, *19*, 252
　Lantana camara, 104, 258
Vitaceae, 143, 220
　Parthenocissus quinquefolia, 143, *144*, 167
　Vitis rotundifolia, 221, *222*, 266
Zamiaceae, 229
　Zamia pumila, 230, 254

www.ingramcontent.com/pod-product-compliance
Lightning Source LLC
Chambersburg PA
CBHW022104150426
43195CB00008B/267